Machine Learning Technology in Biomedical Engineering

Machine Learning Technology in Biomedical Engineering

Editors

Hongqing Yu
Alaa AlZoubi
Yifan Zhao
Hongbo Du

MDPI

Basel • Beijing • Wuhan • Barcelona • Belgrade • Novi Sad • Cluj • Manchester

Editors
Hongqing Yu
University of Derby
Derby
UK

Alaa AlZoubi
University of Derby
Derby
UK

Yifan Zhao
Cranfield University
Cranfield
UK

Hongbo Du
University of Buckingham
Buckingham
UK

Editorial Office
MDPI
St. Alban-Anlage 66
4052 Basel, Switzerland

This is a reprint of articles from the Special Issue published online in the open access journal *Bioengineering* (ISSN 2306-5354) (available at: https://www.mdpi.com/journal/bioengineering/special_issues/ZG3ISDXD72).

For citation purposes, cite each article independently as indicated on the article page online and as indicated below:

Lastname, A.A.; Lastname, B.B. Article Title. *Journal Name* **Year**, *Volume Number*, Page Range.

ISBN 978-3-7258-0803-8 (Hbk)
ISBN 978-3-7258-0804-5 (PDF)
doi.org/10.3390/books978-3-7258-0804-5

Contents

About the Editors

Hongqing Yu

Dr. Hongqing Yu is a prominent Senior Lecturer in Computer Science at the University of Derby, specializing in Data Science and Machine Learning. He holds a PhD in Computer Science from the University of Leicester and has an extensive background in healthcare and biomedical-AI research.

Dr. Yu's contributions to the field of healthcare IT and biomedical research are noteworthy, particularly in the realms of Semantic Web and machine learning. He has been instrumental in a number of EU-funded projects, such as MyHealthAvatar and mEducator, which focus on patient-information systems and medical education, respectively. His role as a principal investigator in projects utilizing machine learning for complex data analysis in aerospace and sports business governance further illustrates his diverse research interests. Dr. Yu has played pivotal roles in numerous high-impact projects, such as the 'NATEP Helping SMEs innovate in aerospace' competition, and has contributed to the development of advanced AI-based solutions in engineering inspection. His expertise extends to supervising PhD students, highlighting his commitment to fostering new talent in the field.

In academia, Dr. Yu is a respected figure, serving as an Editorial Board Member and Guest Editor for various journals in the fields of AI and bioengineering. His publications, particularly on Semantic Web, machine learning, and healthcare IT, demonstrate his significant contribution to the field. Dr. Yu is also actively involved in organizing and participating in international conferences, showcasing his commitment to knowledge sharing and collaboration in the global research community.

Alaa AlZoubi

Dr Alaa AlZoubi is a Senior Lecturer in Computer Science at the University of Derby. He is also an Honorary Senior Research Fellow at the University of Buckingham, a Digital Theme Ambassador for the Digital Theme UK-Ukraine Twinning Initiative and an AI Consultant at Dynamic Edge AI. He holds a PhD in Computer Science from the University of Lincoln, UK. Prior to joining the College of Science and Engineering at the University of Derby, Dr AlZoubi was a Research Fellow in Machine Learning and Computer Vision at The University of Buckingham, as well as a Research Fellow in Computer Vision at Cranfield University. With over 16 years of experience in academia and industry, Dr AlZoubi has made significant contributions to the advancement of machine learning, deep learning, computer vision and software engineering technologies.

Dr AlZoubi is actively involved in research, having contributed to numerous reputable conferences and journals. His research spans a wide range of topics, including Artificial intelligence, AI Decision Explainability, Deep Learning and Computer Vision. He has helped realise several funded research projects, such as automatic cancer detection and land cover analysis, utilising satellite imagery to help with safer landmine surveying. He is currently leading several research projects regarding railway surface inspection and engineering (fleet and aerospace) inspection. Dr AlZoubi is a reviewer in different field related journals, a committee member at different international conferences and a guest editor at Bioengineering.

Yifan Zhao

Professor Yifan Zhao is a Professor of Data Science in the School of Aerospace, Transport and Manufacturing at Cranfield University, UK. He has over 20 years of experience in signal processing, computer vision and utilising artificial intelligence (AI) for degradation assessment and anomaly detection in complex engineering systems. He is dedicated to developing and applying advanced data analysis approaches for solving real-world problems in the sectors of construction ('TRAMS', Innovate UK, 10093011, 'Fuel Coach', BEIS, EEF8037; 'The Learning Camera', Innovate UK, 104794; 'One Source of Truth', Innovate UK, 105881), transport ('CogShift', EPSRC, EP/N012089/1), healthcare ('SecureUltrasound', EPSRC, EP/R013950/1) and supply chain ('RECBIT', Lloyd's Register Foundation (GA\100113). He holds the Royal Academy of Engineering Industrial Fellowship (IF2223B-110) for the development of innovative data-centric solutions aimed at reducing greenhouse gas emissions and fuel consumption and has published more than 200 peer-reviewed journal or conference papers, as well as four books and three patents.

Hongbo Du

Hongbo Du is a Professor of Computing at the University of Buckingham, holding an MPhil in Computing from the University of Essex. His primary areas of research cover big data, data mining and machine learning. His main focus of research in recent years has been on applications of machine learning in medical image analysis, particularly ultrasound images for the diagnosing of cancer. In the past five years, Prof Du has played a leading role as the Director of the TenD Buckingham Research and Development Centre, where several industry-funded research projects have investigated aspects of deep learning in regard to ultrasound image analysis, including image pre-processing, lesion detection, lesion segmentation, recognition and deep learning model decision explanations, with the algorithms regarding cancer sign detections from ultrasound images gaining product certifications from local health regulation authorities in China. He has also worked on collaborative projects involving biomedical image analysis and ovarian ultrasound image analysis with various partners, including the Wellcome Trust Sanger Institute, KU Leaven Hospital and Imperial College Queen Charlotte and Chelsea Hospital. He has published many journals and conference papers in the field of biomedical image analysis and machine learning.

Prof Du has supervised more than 10 PhD theses and 5 MRES dissertations in the past decade. His other areas of research include data stream clustering, sequential data analysis, real-time complex event processing and machine learning for enriching product quality. He has been a regular reviewer for several high impact journals and served on the Editorial Board for PLOS ONE. He is now the deputy director for the Centre of Topological Machine Learning and Innovation at the University of Buckingham.

Preface

The rapid advancement of machine learning and artificial intelligence has inaugurated a new epoch of technological innovation, revolutionizing myriad fields including biomedical engineering. This book, titled "Machine Learning Technology in Biomedical Engineering", stands as a beacon of the extraordinary progress at the nexus of these two disciplines, showcasing the pioneering research and applications poised to redefine healthcare delivery and improve patient outcomes.

Within this compendium, readers will encounter a curated selection of advanced studies that exemplify the confluence of machine learning and biomedical engineering. A noteworthy contribution includes the dense multi-scale graph convolutional network (DMA-GCN) method, tailored for the automatic segmentation of knee-joint cartilage from MR images, presenting a novel paradigm in orthopedic analysis.

Another pivotal study scrutinizes the resilience of leading deep learning architectures, such as ResNet, Swin-Transformer, and MLP-Mixer, against unique distribution shifts in fundus images. This inquiry underscores the imperative of engineering robust models for critical safety applications in medical diagnostics.

The anthology further delineates a sophisticated exploration of gradient-switching-induced potentials during MRI, amalgamating time, frequency, and statistical analysis techniques to enhance MRI image quality and ensure patient safety.

In the domain of non-invasive glucose monitoring, a trailblazing paper unveils an implicit HbA1c model boasting remarkable proficiency in forecasting fasting blood glucose levels by using photoplethysmography data, heralding a new era in diabetes management.

Moreover, this collection introduces a novel machine learning system designed for nutrition education in diabetes mellitus clinics, adept at predicting one-year blood glucose control based on an amalgamation of lifestyle, medication, and clinical data. This innovative modality promises to significantly elevate diabetes care and patient outcomes.

Originally published as a Special Issue in the *Bioengineering* journal, this book is a compilation of cutting-edge research articles that exemplify the remarkable synergy between machine learning and biomedical engineering. One such seminal paper, "AIMS: An Automatic Semantic Machine Learning Microservice Framework to Support Biomedical and Bioengineering Research", signals a paradigm shift in how machine learning can be leveraged to decode the complexities of biomedical data. AIMS exemplifies a quintessential bridge between the theoretical and practical realms, offering a scalable semantic framework that simplifies the intricacies of developing and deploying machine learning solutions in biomedical research. Its significance cannot be overstated, as it not only underscores the potential for machine learning to unravel novel diagnostic and therapeutic pathways but also accentuates the framework's role in facilitating seamless interdisciplinary collaboration, thereby accelerating the pace of innovation in healthcare technologies.

Beyond these examples, this book spans a kaleidoscope of topics, from blockchain-based frameworks for pandemic prevention and control to advanced feature-selection methods for precise disease classification and biomarker identification.

As we navigate through the contents of this book, we are privy to the application of machine learning algorithms in surmounting challenges across the biomedical engineering spectrum. We will delve into groundbreaking approaches that fuse theoretical constructs with practical implementations, nurturing interdisciplinary collaboration and encouraging forward movement in this thrilling area of study.

To the researchers, healthcare practitioners, and students who will immerse themselves in this work, I extend an invitation to approach its contents with an open mind and an inquisitive spirit. Embrace the possibilities at the intersection of machine learning and biomedical engineering. I hope this work inspires you to move beyond the boundaries of what is conceivable in the search for enhanced health and well-being for all.

Hongqing Yu, Alaa AlZoubi, Yifan Zhao, and Hongbo Du
Editors

bioengineering

MDPI

Article

AIMS: An Automatic Semantic Machine Learning Microservice Framework to Support Biomedical and Bioengineering Research

Hong Qing Yu *,†, Sam O'Neill † and Ali Kermanizadeh

School of Computing and Human Sciences Research Centre, University of Derby, Derby DE22 3AW, UK; s.oneill@derby.ac.uk (S.O.); a.kermanizadeh@derby.ac.uk (A.K.)
* Correspondence: h.yu@derby.ac.uk
† These authors contributed equally to this work.

Abstract: The fusion of machine learning and biomedical research offers novel ways to understand, diagnose, and treat various health conditions. However, the complexities of biomedical data, coupled with the intricate process of developing and deploying machine learning solutions, often pose significant challenges to researchers in these fields. Our pivotal achievement in this research is the introduction of the Automatic Semantic Machine Learning Microservice (AIMS) framework. AIMS addresses these challenges by automating various stages of the machine learning pipeline, with a particular emphasis on the ontology of machine learning services tailored to the biomedical domain. This ontology encompasses everything from task representation, service modeling, and knowledge acquisition to knowledge reasoning and the establishment of a self-supervised learning policy. Our framework has been crafted to prioritize model interpretability, integrate domain knowledge effortlessly, and handle biomedical data with efficiency. Additionally, AIMS boasts a distinctive feature: it leverages self-supervised knowledge learning through reinforcement learning techniques, paired with an ontology based policy recording schema. This enables it to autonomously generate, fine-tune, and continually adapt to machine learning models, especially when faced with new tasks and data. Our work has two standout contributions demonstrating that machine learning processes in the biomedical domain can be automated, while integrating a rich domain knowledge base and providing a way for machines to have self-learning ability, ensuring they handle new tasks effectively. To showcase AIMS in action, we have highlighted its prowess in three case studies of biomedical tasks. These examples emphasize how our framework can simplify research routines, uplift the caliber of scientific exploration, and set the stage for notable advances

Keywords: AI automation; biomedical; machine learning; microservices; knowledge graph; semantic web services (SWS)

Citation: Yu, H.Q.; O'Neill, S.; Kermanizadeh, A. AIMS: An Automatic Semantic Machine Learning Microservice Framework to Support Biomedical and Bioengineering Research. *Bioengineering* **2023**, *10*, 1134. https://doi.org/10.3390/bioengineering10101134

Academic Editor: Yunfeng Wu

Received: 30 August 2023
Revised: 21 September 2023
Accepted: 25 September 2023
Published: 27 September 2023

1. Introduction

The fusion of machine learning and biomedical and bioengineering research has brought a paradigm shift in the way we understand, diagnose, and treat an array of health conditions. With rapid advancements in technology and an influx of high-dimensional data, the role of machine learning (ML), and especially AutoML, has become central to the process of knowledge discovery in this field [1]. Providing a machine learning solution often creates a burden for biomedical and bioengineering researchers, who must seek additional support to develop or test different tools for each step of their study. The burgeoning complexity of biomedical research underscores the need for the automatic generation and optimization of machine learning models that can keep pace with this data-driven evolution [2]. The impetus for our research stemmed from challenges faced by our university's biomedical research group. They grappled with integrating a myriad of tools, algorithms, and domain-specific knowledge in their investigative pursuits. We responded by devising a framework tailored for these exact scenarios. This methodology represents an innovative approach, unprecedented in its application across any other research domain. Such distinctiveness makes our contribution particularly fitting for this special issue on machine learning technology in biomedical engineering. In this paper, we investigate the novel approach of applying an advanced ontology and context enforcement learning approach that can support the automation of a machine learning process for biomedical and bioengineering research. This work has great potential to be applied to other domain areas, but the background ontology requires updating according to the domain knowledge.

The AutoML frameworks provide a robust solution to this rising demand. They offer end-to-end pipelines that encompass all necessary steps from data preprocessing to hyperparameter tuning and model evaluation, automating labor-intensive and error-prone manual tasks [3]. By significantly reducing the time taken for the model development process, they allow researchers to focus on interpreting and applying results, thus accelerating the pace of discovery in the biomedical and bioengineering field.

Despite the undeniable potential of AutoML, several gaps and challenges persist in its implementation in biomedical and bioengineering research. First, biomedical data, with their unique characteristics, including high dimensionality, heterogeneity, and inherent noise, require specialized preprocessing and analytical approaches. Current general-purpose AutoML frameworks may not adequately address these needs. Second, these frameworks often lack interpretability, a crucial requirement in the medical field, where understanding the decision-making process of a model is as essential as its predictive accuracy [4]. Finally, integrating domain knowledge into the AutoML process remains an open challenge, although it could greatly improve the quality of models generated and the applicability of their predictions [5]. Addressing these challenges requires a novel framework that enables machines to learn domain-specific knowledge and apply this knowledge to automate decisions in the AutoML process.

This paper introduces an Automatic Semantic Machine Learning Microservice framework designed to bridge these gaps. We refer to each microservice in our framework as AIMS, and these should be implemented based on domain-specific knowledge. It is tailored to the specific needs of biomedical and bioengineering research and places emphasis on enhancing model interpretability, incorporating domain knowledge, and handling the intricacies of biomedical data. Our proposed framework aims to streamline the research process, augment the quality of scientific exploration, and provide a foundation for significant self-learning AutoML in biomedical research. Additionally, three case studies are tested and discussed at the end.

2. Related Work, Limitations, and Technology Background

In the work of the biomedical and bioengineering research community, the most important challenge is to search for different tools that work on different datasets and tasks. Our research also begins by examining existing automation technologies and tools.

2.1. Related Work and Current Limitations

There is a multitude of AI tools currently available to aid biomedical research, and various automation frameworks have emerged to streamline the process. For instance, machine learning platforms like Google's TensorFlow [6] and scikit-learn [7] have been widely utilized in biomedical research for tasks such as image analysis, genomics, and drug discovery.

Google's AutoML [8], TPOT (a tree-based pipeline optimization tool) [9], and H2O's AutoML [10,11] are some of the popular AutoML tools used for automating the machine learning pipeline. These platforms optimize the process by automating tasks like data preprocessing, feature selection, model selection, and hyperparameter tuning, which are traditionally labor-intensive and error-prone. AutoML offers expedited results, bypassing much of the manual work involved in traditional machine learning, which is especially advantageous for prototype testing or gauging initial user reactions to AI applications. Moreover, AutoML solutions are less prone to becoming obsolete, as they can stay updated with rapid advancements in AI technology, largely due to the investment capacity of major tech vendors. Additionally, AutoML platforms, being hosted solutions, reduce the overhead of building surrounding infrastructure. TPOT aims to simplify the construction of ML pipelines by merging a versatile expression tree depiction of these pipelines with random search techniques like genetic programming. It leverages the scikit-learn library in Python as its foundation for machine learning functionalities. H2O's AutoML streamlines the machine learning process by autonomously training and fine-tuning various models within a time frame set by the user. Additionally, H2O incorporates several model interpretability techniques applicable to both AutoML collections and distinct models, such as the leader model. These explanations can be effortlessly produced with a singular function, offering an intuitive means to probe and elucidate the AutoML models.

Other than these general-purpose tools, there are also specialized AI platforms tailored for biomedical research. DeepChem [12], for instance, is a machine learning library specifically designed for drug discovery and toxicology, offering specialized features not available in general-purpose libraries.

However, while these tools have made significant strides in advancing biomedical research, there are several limitations associated with their use.

General-purpose ML and AutoML tools, such as TensorFlow and Google's AutoML, are not specifically designed for handling the unique characteristics of biomedical data, such as high dimensionality, heterogeneity, and inherent noise. This often necessitates significant manual preprocessing before data can be fed into these tools [13].

Furthermore, these tools often lack interpretability, an essential requirement in biomedical research, where understanding the decision-making process of a model is as important as its predictive accuracy [4].

While specialized tools like DeepChem offer features tailored for biomedical applications, they do not cover the entire spectrum of biomedical research and are limited in their scope. Additionally, the automatic integration of domain knowledge into the machine learning process is an ongoing challenge and is not well addressed by current tools [4].

Therefore, there are many recent discussions on multiple biomedical task handling with self-learning, self-optimizations, and self-configuration processes, such as [14], focusing on data science processing automation with optimization, and [3], focusing on feature selection and model training.

2.2. Multiple-Task AI System Research

Making the system automatic by learning the solution knowledge about the different tasks is also challenging. Industrial AI leading research groups such as Google AI and Meta AI understood that data-driven AI technologies have issues with performing complex tasks, for example, creating human conversations with contextual understanding or detecting early signs of disease from images. In addition, data-driven AI is resource-intensive and suffers from algorithm bias [15]. Thus, the multiple-task-enabled AI systems with a knowledge-driven approach present a pathway toward a solution to these problems. Why is it thought that a knowledge-driven approach is necessary and crucial for a multiple-task system? There are two reasons:

- The information acquired from different tasks may present value that can be used as the basis to build new ML models for new tasks without requiring high-cost processing to recapture the same feature characteristics.
- Updating knowledge through validation is a relatively consistent process that will be less prone to bias from noisy data.

Upon completion of this research, two new ideas from Google and Meta were published.

Google presents an experimental process based on knowledge mutation [16,17]. Here, knowledge refers to base neural network transformers. To begin with, the experimental environment contains transformers which can work on different tasks (different image datasets for the classification problems). Then, when a new task arrives, the most related transformer will be triggered to perform a mutation process. The mutation process can edit the base model by inserting a new layer, removing a layer, or doing both according to the performance optimization. In the end, a new mutated adapter is created to enable dealing with a similar task next time. Whenever a new task with a new dataset arrives, the mutation process is executed based on the latest mutated model.

The Meta research group presents a world model approach to acquiring knowledge, very much in the spirit of actor–critic reinforcement learning [18]. The system architecture is a combination of smaller modules—configurator, perception, world model, cost, short-term memory, and actor—that feed into each other. The world model module is responsible for maintaining a model of the world that can then be used to both estimate missing information about the world and predict plausible future states of the world. The perception module will receive signals to estimate the current state of the world and, for a given task, the configurator module will have trained the perception module to extrapolate the relevant signal information. Then, in combination, the perception, world model, cost, short-term memory, and actor modules feed into the configurator module, which configures the other modules to fulfill the goals of the task. Finally, the actor module is handed the optimal action to perform as an action. This has an effect on the real world which the perception module can then capture, which in turn triggers the process to repeat. That is, each action will produce a piece of state-changing knowledge feedback to the world model for continuous learning.

Both Google and Meta's visions derive from the previous hyperparameter-optimization-based AutoML processes [19]. For example, AutoKeras [20], a neural architecture autosearch framework, is proposed to perform network morphism guided by Bayesian optimization and utilizing a tree-structured acquisition function optimization algorithm. The searching framework selects the most promising Keras implemented NN for a given dataset.

The above experimental results show improvements in tackling complex AI tasks and possible pathways toward human-level AI systems. However, there are two main limitations:

- The knowledge definition is too narrow and only uses the generated neural network as the knowledge limits the capability of recording all valuable outcomes through the learning experience.
- There is no unified knowledge representation structure for knowledge inference (machine thinking).

Do we already have a knowledge representation framework from our AI research over the past 70 years? The answer is yes.

2.3. *Knowledge Representation and Reasoning*

Knowledge representation and reasoning (KRR) are always the core research areas in AI systems [21]. Knowledge representation and reasoning (KRR) is a core area of artificial intelligence (AI) that deals with how to symbolically represent information in a way that a computer system can use to reason about the world. This involves understanding and emulating human-like thinking and the ability to make deductions, inferences, or predictions. KRR aims to enable machines to represent knowledge in a manner that they can reason with, as humans do. Here are the main components:

Knowledge representation (KR): This is about how to store, retrieve, and modify knowledge in an intelligent system. Various paradigms like semantic networks, frames, rules, and ontologies have been developed for this purpose.

Semantic networks: Graph-based structures used to represent knowledge, where nodes represent concepts and edges represent relationships between concepts.

Frames: Data structures for representing stereotypical situations. They contain attributes (or slots) and associated values.

Rules: Represent knowledge in terms of if–then statements.

Ontology: Define a set of representational primitives with which to model a domain of knowledge. Ontologies are used in modern AI applications, especially in the semantic web.

Knowledge reasoning: This is about using the stored knowledge to draw conclusions, make decisions, or infer new knowledge.

The knowledge-enhanced machine learning approach attracts less attention than the data-driven approaches. However, KRR is still key in developing the future generation of AI system development [22]; even Deep Neural Networks (DNNs) can create KRR, just in a different form [23]. In our vision, KRR should not only extract knowledge from data but also learn knowledge from system actions that can support the reasoning process. Knowledge reasoning can be seen as the fundamental building block that allows machines to simulate humankind's thinking and decision making [24]. With generations of development on KRR, the current most promising approach is the knowledge graph (KG) [24], derived from the semantic web [25] community. A knowledge graph has two layers of representation structure: 1. predefined ontology and vocabularies and 2. instances of triple statements (e.g., dog is animal, where the dog is an instance and is is a predict, while animal is a concept vocabulary defined in the ontology). The reasoning part is to apply the logical side of the ontology, such as description logic (e.g., Is the dog an animal? The reasoning result is 'yes') [26]. There have been many complex types of ontologies developed in the last decade to solve different KRR problems and applications. The most important development of ontology-driven reasoning is to encode dynamic uncertainty [27], probability [28], and causality [29]. Therefore, the KG-based KRR framework can be applied to implement our proposed vision.

2.4. Services and Machine Learning Ontologies

The web services community has researched autoconfiguration or service composition for many years by applying a variety of dynamic integration methods. There are two trends in service composition research:

- Directly extracting the services description file (e.g., WSDL) and Quality of Services (QoS) into a mathematical model with a logical framework for composing services such as a linear logic approach [30] and genetic algorithms [31,32]. The major limitation is that there are no formal specifications for modeling and reasoning. Therefore, the processes are mostly hard-coded to match the logic framework.
- The other trend is to apply semantic web standards for semantically encoding service descriptions and their QoS properties (semantic web services (SWS)) [33,34]. The main benefit is that semantic annotation has an embedded logical reasoning framework to deal with composition tasks.

On the one hand, the semantic web services (SWSs) trend has greater strength for integrating the KRR approach with the same semantic infrastructure and reasoning logic. Currently, there are three standards of OWL-S: composition-oriented ontology, WSMO (task–goal matching-oriented ontology), and WSDL-S (invocation-oriented ontology). On the other hand, there are two differences between our vision's microservice and normal SWSs. The first one is that AIMS has simpler input and output requirements to perform an efficient composition process. The other is that the purpose of each microservice is to deal with data analytic or machine learning tasks. Therefore, the AIMS ontology needs to be defined by modifying current machine learning ontology standards. Researchers have realized that there is a need to have a machine learning ontology, and some recent proposals in this domain are the Machine Learning Schema and Ontologies (MLSO), which introduces twenty-two top-layer concepts and four categories of lower-layer vocabularies (the detailed ontology design is in [35]), and the Machine Learning Ontology (MLO), which proposes to describe machine learning algorithms with seven top-layer concepts of Algorithm, Application, Dependencies, Dictionary, Frameworks, Involved, and MLTypes [36].

The existing ontology and schema provide a foundational base that can be integrated and augmented to define a more comprehensive schema for generating automotive AI solutions in the biomedical domain. The primary enhancement required is to effectively present the knowledge acquired from each task. Additionally, a self-learning policy is essential to aid machines in comprehending the task context and devising the most optimal pathway to offer a solution.

2.5. Generative AI

Recently, generative AI technology, such as ChatGPT and its associated APIs, has marked a significant advancement in AI research. These technologies are primarily designed to engage in text-based conversations, providing solutions to queries and problems in a natural, human-like manner. This form of AI has shown tremendous utility in diverse fields, from customer service to education, demonstrating its versatility.

However, when applied to more complex domains like biomedical research, there are notable limitations. Specifically, the ability of these models to generate code or automated solutions for multistep biomedical problems is limited. The key issue lies in the representation and understanding of data tokens within these problem spaces. In biomedical research, data tokens can represent complex and highly specific biological or medical entities, procedures, or relationships, which can be challenging for AI models to comprehend.

Generative AI models like ChatGPT operate best when dealing with structured data and clear-cut problem domains. Yet, biomedical research often involves dealing with unstructured or semistructured data, highly domain-specific language and concepts, and complex multistep processes.

Another significant challenge for generative AI, particularly in highly specialized fields such as biomedical research, is the integration and expansion of domain-specific human knowledge within the existing large language model.

Generative AI models are usually trained on extensive and diverse datasets, covering a broad range of topics and languages. As a result, they can effectively generate text that mimics human language in many situations. However, these models typically lack the ability to learn continuously or integrate new knowledge once they have been trained. Their knowledge is essentially frozen at the point of their last training update.

This limitation becomes particularly problematic when attempting to apply these models in rapidly advancing fields such as biomedical research, where new discoveries and innovations continually push the boundaries of existing knowledge. As the model cannot natively integrate this new information, it struggles to provide up-to-date and accurate solutions to complex, domain-specific problems. This limitation also extends to learning from user interactions over time, a process which could theoretically allow the model to fine-tune its responses and become more accurate.

Furthermore, the vast and generalized knowledge base of these models can be a double-edged sword. While it allows them to engage with a wide variety of topics, it can also lead to dilution of specialist, domain-specific knowledge. The models may struggle to produce in-depth, nuanced responses to specialized queries due to the sheer breadth of their training data.

In summary, while generative AI has shown significant promise, its limitations in integrating and extending domain-specific human knowledge, coupled with its inability to learn continuously, present considerable challenges for its application in specialized fields like biomedical research. Overcoming these challenges will require novel approaches to model training and updating, making it an exciting area for future AI research and development. A potential strategy could involve using pretrained, domain-specific transformers as a base model. This would facilitate the use of customized small research datasets to efficiently produce a high-quality model. However, this approach necessitates a base model framework to select the most suitable transformer effectively.

2.6. The Gaps

By reviewing the current state of the art, we found that there are research gaps remaining to achieve our goal:

- Self-supervised knowledge generation during the machine learning process and solution creation: In the past, knowledge generation system mainly referred to expert systems that acquire knowledge from human expertise or systems that transform existing knowledge from one presentation to the other. Enabling the understanding of common knowledge in the biomedical domain is crucial. Reference [29] presents an automatic process of disease causality knowledge generation from HTML-text documents. However, it still does not fully address the problem of how to automatically learn valuable knowledge from the whole task–solution–evaluation machine learning life cycle. Considering human-level intelligence, we always learn either directly from problem-solving or indirectly through other human expertise (e.g., reading a book or watching a video), or a combination of both (e.g., reflecting on the opinions of others).
- Provisioning a knowledge-guided auto-ML solution: In contrast to the first gap, there are no significant research works on using knowledge to assist in providing an AI solution. Again, compared with human-level intelligence, we always try to apply acquired knowledge or knowledge-based reasoning to solve a problem. We can consider that the transformer process [37] is a step forward in this direction. We can treat well-trained AI models as a type of knowledge to apply to different tasks in a similar problem domain. However, there is still no defined framework that can specify what knowledge is required and how to use the knowledge to find a solution to new tasks [16].

3. The Framework Architecture

Figure 1 represents our vision of self-supervised knowledge learning with the AIMS engineering approach. The left part of the Figure 1A presents the initial settings of the intelligent environment. The initial environment only contains default AIMS information, such as purposes, I/O requirements, and invokable URI (detailed AIMS metadata ontology is introduced in the next section). However, the initial settings are ready to perform four things:

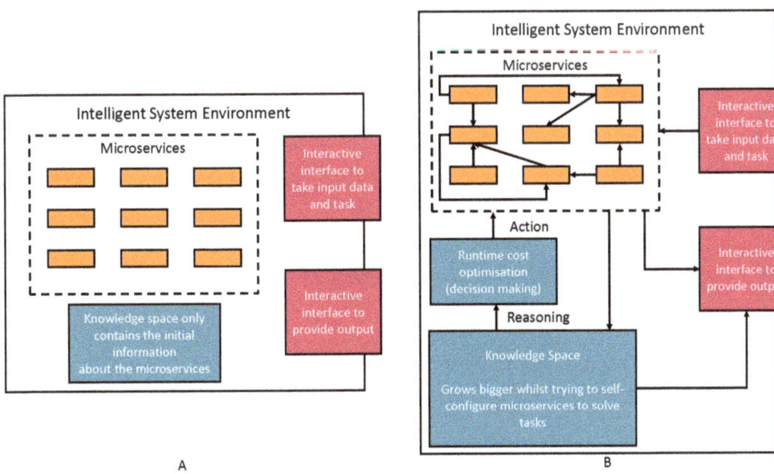

Figure 1. The vision of the self-knowledge learning approach with semantic ML microservices. (**A**) shows the initial environment of the framework without any knowledge learning. (**B**) shows the knowledge growing after learning from the actions that tried to provide an solution to the given tasks.

- Registering new AIMSs (Automatic Semantic Machine Learning Microservices) from outside the environment. The registration process is through the interactive interface according to the defined microservice ontology (see Figure 2). Therefore, human involvement in machine learning microservice engineering is a core part of this vision, which defines humans as educators to teach basic skills and capabilities to deal with different tasks. Then, the environment will reuse these skills and capabilities to acquire knowledge. The knowledge will provide powerful reasoning sources to independently deal with complex tasks, decision making, and creating new pipelines. Specific to the biomedical research, the registration ontology can refer to biomedical engineering ontologies, including Disease Ontology, Foundational Model of Anatomy (FMA), Human Phenotype Ontology (HPO), and many others [38].

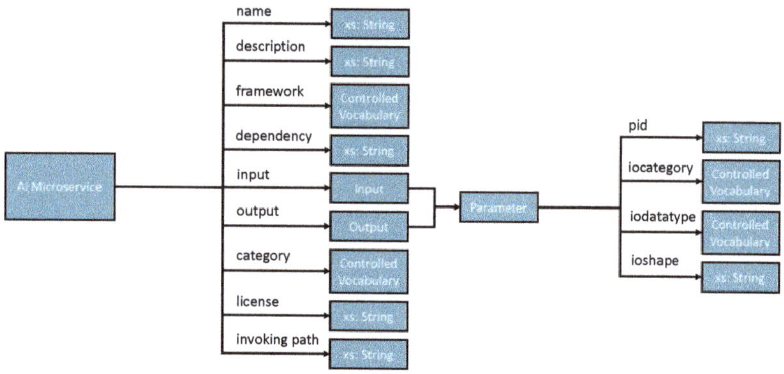

Figure 2. AI Microservice Registration Ontology.

- Taking tasks with a variety of inputs, such as CSV data files, images, text, and audio data: The environment autoconfigures on the default AIMSs and provides solutions to the tasks. The success or failure outcome will be recorded as knowledge. Microservice autoconfiguration refers to the automated setup and configuration of individual microservices in a pipeline to serve a machine learning task in our context. The microservice human engineering process will start if there are no suitable AIMSs to deal with the task.
- The environment can compose multiple AIMSs to complete a task if one single microservice cannot achieve it.
- The environment can start learning, representing, and storing knowledge in the knowledge space as knowledge graph data. The knowledge is derived from processing input data, the autoconfiguration process, and task outcomes. The knowledge size will increase and thus provide better optimizations, autoconfiguration, and feedback to the system user.

To realize the vision presented in Figure 1, we discuss the related existing technologies and their research outcomes that can be adapted into our research next.

4. Self-Supervised Knowledge Learning for Solution Generation

The self-supervised knowledge learning approach involves three types of autoconfiguration transfer learning methods. Figure 3 presents the overall learning framework.

The first method is knowledge space searching and transferring: A task with a dataset (referred to as a task-context) arrives, and there is no previous knowledge related to the task-context. Therefore, the knowledge space will be searched to try to find a possible microservice that can match the context to complete the task or search for a pipeline (workflow) that contains multiple I/O compatible AIMSs together towards the best and successful completion which can be optimized. The task context and the optimized solution are recorded as task input and output knowledge.

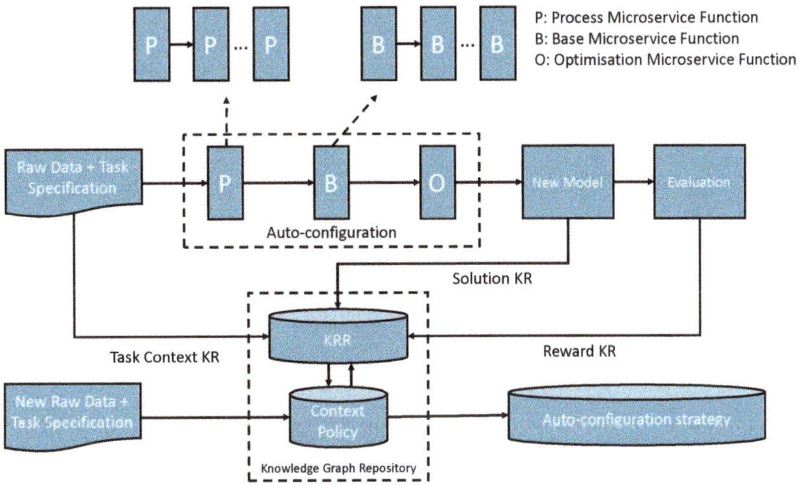

Figure 3. The vision of the self-knowledge learning approach with AI microservices.

The evaluation will generate rewards for the policy knowledge space (we demo a detailed process in the sections Experimental Implementation and Scenario Evaluation and Lesson Learned). In addition, the knowledge learned from the process will be recorded to update the world knowledge space.

The second one is the mutation of a previously generated context-matched solution (a composition transfer learning process; we demo a detail process in the section Scenario Evaluation and Lesson Learned). If the new task context matches with a previously recorded task context in the knowledge space, then the previous solution will be loaded to adapt to the new datasets and the optimization process. Finally, a new mutated solution is created and recorded as new knowledge with the new evaluation rewards and world knowledge of the KRR environment.

The third one is the continuous learning mutation method based on the reinforcement learning approach. With the growth of the KRR statements, the automutation will take place using world knowledge to retrain the solutions according to the rewards. The third learning method takes place offline only but continues carrying out an update when KRR is updated.

5. Experimental Implementation

Figure 4 shows a three-layer implementation of the vision. This structure reflects to our vision that AI system should have three major capabilities of learning knowledge, reasoning (thinking), and reacting to the problem. The figure also shows how these layers map to the automotive solution provisioning process.

- The request layer takes tasks and inputs from AI applications to trigger the solution searching and self-learning processes. Task context is semantically encoded to enable starting the policy knowledge to explore the environment for learning, creating, or finding solutions.
- The reasoning layer takes the request layer's semantic reasoning tasks for semantic matching, reasoning, and performing the reinforcement learning mechanism. Finally, the policy will be recorded in the knowledge graph layer. In addition, the newly added AIMSs are registered to the environment with semantic annotations through knowledge registration and generation components.
- The knowledge graph layer remembers the knowledge data in the knowledge graph triple store based on different types of knowledge schemata.

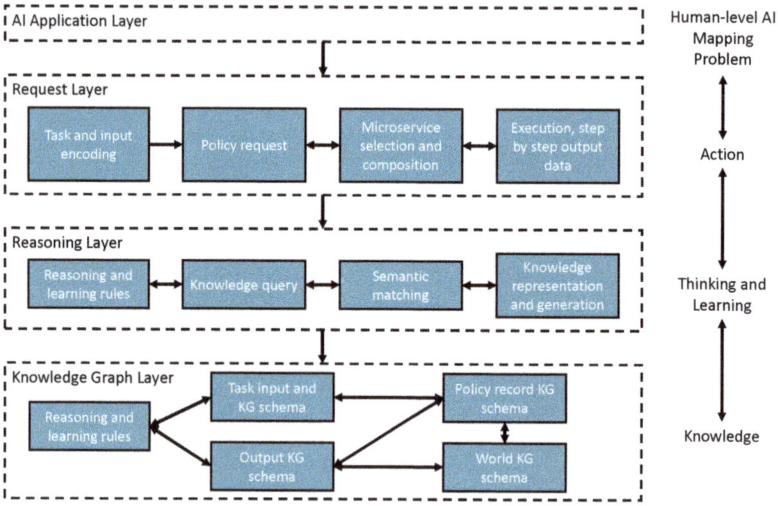

Figure 4. Self-supervised knowledge learning and reasoning framework design.

5.1. Knowledge Ontology Implementation

AIMS registration ontology defines nine parameters (see Figure 2).

- Name— Must be no duplication in the system, and the registration process will check the name's legibility.
- Description—A short presentation of the AIMS for human understanding.
- Framework—Indicates the programming framework used to develop the AIMS. Normally, it should be just one framework as AIMS is designed to be decoupled and ideally single responsibility.
- Dependency—Describes the required programming libraries that need to be preconfigured to enable the AIMS to work. The schema includes id, library install port URI and version.
- Input and output—Specifies the parameters that should be in the input and output messages.
- Category—Tells what AI-related domain the ms works on, such as supervised classification, unsupervised clustering, image classification base model, and more.
- License—Identifies the use conditions and copyright of the AIMS.
- Invoke path—Contains the portal for accessing the AIMS. The path can be a local path or URI of a restful API.

Each given task triggers a context knowledge creation that collects knowledge of the following:

- The type of input data—A controlled variable that majorly includes normal dataset (e.g., tableau data stored in a CSV file, image, or text).
- Task domain—Free text to record the specific application domain.
- Desire output type—Records the output required to complete the task successfully.
- The parameters—The dataset or data presents the initial characters of the input data. For example, the number of columns and column names of the tableau data will be remembered as part of the context knowledge of the task.

The output of the performed task can be categorized into two types, failure and success. Both failure and success need to update the policy knowledge link to the task-input context. Failure has no solution registered to the knowledge but records which AIMSs have been successfully invoked (can be an empty list) until the step that cannot continue going further. So, the failure experience will tell the system administrators (people) what AIMS(s) are required to create a solution. The success registers the solution location and changes the policy with the reward value. If the solution contains a workflow of composed AIMSs, then the workflow will also be registered as knowledge with the normalized rewards for each of the AIMSs.

The policy ontology is designed as follows:

- Policy context—Links to a task-context.
- Policy state—1 is success and 0 is failure, the binary state only presents whether the whole workflow is work or not but inside the workflow context that shows the continuous measure of individual component's potential contribution toward to success in other possible solutions (see Figure 5).
- Solution iloc—The location where the solution can be loaded and executed.
- Workflow—Presents a pipeline solution that composes multiple AIMSs.
- Solution reward—The reward value stored for the policy that can be the recommended guidance for supporting the creation of a new task solution.

```
ns1:68819464-461d-4253-a1c0-ddd27c2bedd2 ns1:context ns1:49ce4467-499c-42b5

ns1:49ce4467-499c-42b5-b4da-b325d393a17a ns1:policy_state "0"@en ;
    ns1:solution ns1:305dab4e-b4e9-4b5a-970f-63e7745b11db ;
    ns1:workflow [ [ ns1:reward "0.3333"@en ;
                     ns1:wf_id "0"@en ;
                     ns1:wf_iloc "loadnormalpddata"@en ],
                   [ ns1:reward "0.3333"@en ;
                     ns1:wf_id "1"@en ;
                     ns1:wf_iloc "spliting"@en ],
                   [ns1:reward "0"@en ;
                     ns1:wf_id "3"@en ;
                     ns1:wf_iloc "modelevaluation"@en ],
                   [ns1:reward "0.3333"@en ;
                     ns1:wf_id "2"@en ;
                     ns1:wf_iloc "pipelinemodels"@en ],
                 ] .
```

Left: failure scenario

```
ns1:d1e83aee-7fd0-44a9-bd9a-755312e0e397 ns1:context ns1:49ce4467-499c-42b5

ns1:49ce4467-499c-42b5-b4da-b325d393a17a ns1:policy_state "1"@en ;
    ns1:solution ns1:5b128126-86b2-47ae-8517-5c13c5114fd9 ;
    ns1:workflow [ [ ns1:reward "0.1875"@en ;
                     ns1:wf_id "1"@en ;
                     ns1:wf_iloc "spliting"@en ],
                   [ ns1:reward "0.0625"@en ;
                     ns1:wf_id "3"@en ;
                     ns1:wf_iloc "featuremodelevaluation"@en ],
                   [ ns1:reward "0.125"@en ;
                     ns1:wf_id "2"@en ;
                     ns1:wf_iloc "pipelinemodels"@en ],
                   [ ns1:reward "0.25"@en ;
                     ns1:wf_id "0"@en ;
                     ns1:wf_iloc "loadnormalpddata"@en ] ] .

ns1:5b128126-86b2-47ae-8517-5c13c5114fd9 ns1:reward "1"@en ;
    ns1:s_iloc "KGLayer/models/2c432a85-ac57-4a4e-86bb-54d280400aad.gz"@en
```

Right: success scenario

Figure 5. Reinforcement learning policy generation for the knowledge layer.

The can be rewarded in a failure pipeline if the individual step is invoked and running successfully. Therefore, the rewarded micorservice can be reused when searching for the alternative success composition solution. Figure 5 shows that failure workflow at runtime provide one of three rewards to the successful individual microservices, but no final solution model is created comparing it with the second searching which created a successful workflow pipeline and recalculated rewards to all four microservices. The successful workflow will reward the total reward value 1 divided by the number of microservices (n) involved (n—order number).

The world knowledge ontology presents the facts learned from the task solution creation process and outputs. There are three types of world knowledge recorded in the current environment:

- Feature optimization outcomes—The features selected in the optimization are valuable, and these features will be reused to create a classification model if the new dataset features are the same.
- Answers for a certain text topic—A generated text answer for a question. The answer quality will be reported as a reward value feedback from humans back to the policy knowledge.
- Image RBG vectors—Map to a classification label. The reward process is the same as the answers.

More world knowledge can be expanded in the environment. By having these commonsense and policy records, reinforcement can be performed to improve the solution accuracy incrementally.

5.2. Environment Initialization

The experiment environment is developed by Python in a local single-computer environment. We simplified the AIMS as a .py module in the environment to be invoked and registered. We initialized the environment with three types of AIMSs:

1. Data processing AIMSs that include CSV files to a Panda service, data training and split services, data quality control services, data normalization services, image process services, and data quality control services.
2. ML AIMSs that include clustering services, classification services, GPT-neo-1.3B text generation services [39,40], ViT image classification transformers [41], and Seanborn visualization services.
3. RFECV optimization services.

6. Scenario Evaluation and Lessons Learned

6.1. Heart Disease Classification Scenario

Figure 6 presents one of our use-case scenarios in the medical domain. The task context is as follows:

- Input: 335 clinical CSV heart disease files labeled 0 (no disease) and 1 (confirmed disease).
- Domain: Medical.
- Desire output: An optimized classification pipeline model.

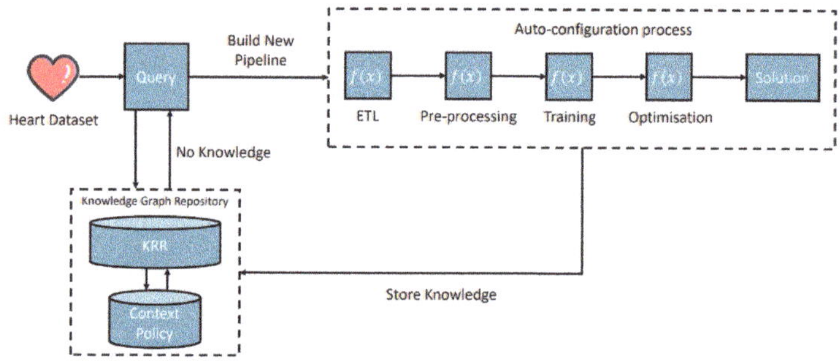

Figure 6. Scenario 1—Heart disease classification solution building and knowledge learning process.

Additionally, the columns of the data are as follows:

- Age: The person's age in years.
- Sex: The person's sex (1 = male, 0 = female).
- cp: Chest pain type—value 0: asymptomation, value 1: atypical angina, value 2: nonanginal pain, value 3: typical angina.
- trestbps: The person's resting blood pressure (mm Hg on admission to the hospital).
- chol: The person's cholesterol measurement in mg/dL.
- fbs: The person's fasting blood sugar (>120 mg/dL, 1 = true; 0 = false).
- restecg: Resting electrocardiographic results.
- thalach: The person's maximum heart rate achieved.
- exang: Exercise-induced angina (1 = yes; 0 = no).
- oldpeak: ST depression induced by exercise relative to rest (ST relates to positions on the ECG plot).
- Slope: The slope of the peak exercise ST segment—0: downsloping; 1: flat; 2: upsloping.
- ca: The number of major vessels (0–3).
- thal: A blood disorder called thalassemia.
- target Heart disease (1 = no, 0 = yes).

The application domain is medical, and the desired output is an optimized classification pipeline model. Figure 6 illustrates the process. The procedure commences with the input of a CSV file containing the Heart Disease dataset, accompanied by a query pertaining to the generation of a predictive model. Following this, the framework attempts to retrieve relevant knowledge via the request layer and its associated functions, as shown in Figure 4. Given the absence of pre-existing knowledge, the system initiates a reasoning process through the reasoning layer. This layer, utilizing the knowledge representation of existing microservices, crafts a workflow. The formulated workflow incorporates four AI microservices: data loading from the CSV, data partitioning, creation of a classification pipeline, and optimization. The culmination of this process results in a model boasting an accuracy rate of 96.8%. Throughout the procedure, various knowledge components are assimilated and documented within the system's environment. The unique insight derived from the general knowledge context is that eight features are deemed more significant than other columns in determining the classification results. These features are sex, cp, thalach, exang, oldpeak, slope, ca, and thal. The reason these columns are highlighted as the most crucial is because the optimization microservice identified them in generating the most accurate model.

6.2. Parkinson Disease Classification Scenario

The second task context is as follows:

- Input: CSV Parkinson's disease clinical example data with labels 0 (no disease) and 1 (confirmed disease).
- Domain: Medical.
- Desired output: An optimized classification pipeline model.

Figure 7 depicts the scenario in which a similar task of classifying Parkinson's disease is fed in; the framework starts searching for a solution. As the system environment has preknowledge, gained through the previous heart disease classification, and since the only difference is the dataset when compared with the heart disease classification context, the framework can use the classification pipeline and retrain it to be optimized for the new dataset. We can call this process a composition transfer learning process. The novelty is that the system environment can solve different tasks by applying contextual knowledge of the problem. Thus, the framework can automatically deal with all types of data if the required models are semantically registered in the framework. Through this composition transfer learning, the whole automatic pipeline can produce a 94.6% accurate model.

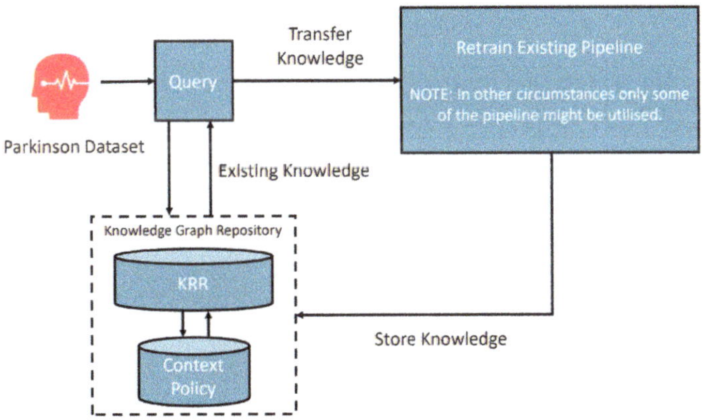

Figure 7. Scenario 2—Parkinson's disease pipeline transfer classification process.

6.3. A Complex Scenario: Mouse Brain Single-Cell RNASeq Downstream Analysis

In this section, we use a clustering analysis case study to highlight how the proposed framework can solve a real-world downstream single-cell data analysis task. The clustering analysis of single-cell data offers a powerful tool for a myriad of applications, ranging from understanding basic biological processes to the development of clinical strategies for treating diseases. The clustering analysis task works on a mouse brain single-cell RNASeq dataset. The dataset is publicly available through a workshop tutorial at [42]. There are five sequential processing and analysis steps:

1. Data semantic transforming and loading: For instance, applying AnnData structure [43], where AnnData stores observations (samples) of variables/features in the rows of a matrix (see Figure 8).
2. Data quality control: This aims to find and remove the poor-quality cell observation data which were not detected in the previous processing of the raw data. The low-quality cell data may potentially introduce analysis noise and obscure the biological signals of interest in the downstream analysis.
3. Data normalization: Dimensionality reduction and scaling of the data. Biologically, dimensional reduction is valuable and appropriate since cells respond to their environment by turning on regulatory programs that result in the expression of modules of genes. As a result, gene expression displays structured coexpression, and dimensionality reduction by the algorithm such as principle component analysis can group these covarying genes into principle components, ordered by how much variation they explained.
4. Data feature embedding: Further dimensionality reduction using advanced algorithms, such as t-SNE and UMAP. They are powerful tools for visualizing and understanding big and high-dimensional datasets.
5. Clustering analysis: Groups cells into different clusters based on the embedded features.

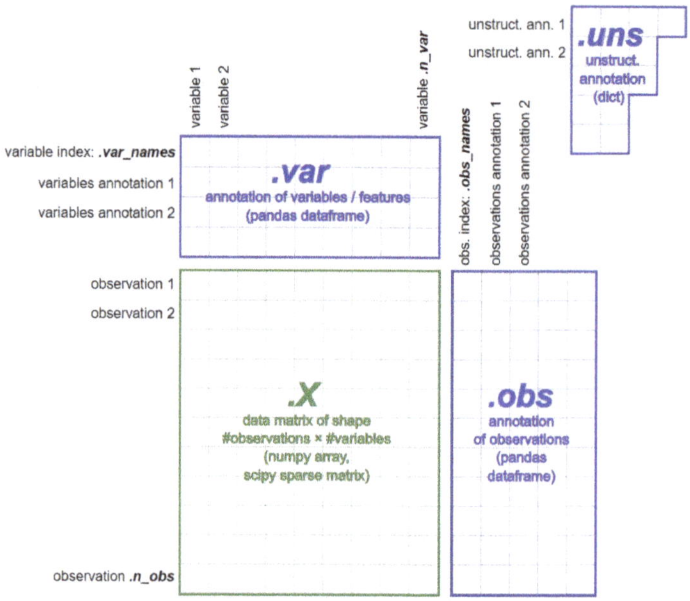

Figure 8. AnnData Structure.

Based on the above five steps, we developed four extra microservices, which include AnnData loading, two feature embedding services (t-SNE and UMAP), and clustering services (Louvain graphical clustering algorithms). The other existing microservices which can be involved should be different types of normalization (PCA or CPM algorithm) and K-mean clustering algorithms.

The microservices were semantically registered into the framework through the interface. Figure 9 depicts an example of a quality control microservice semantic description in the knowledge graph repository.

```
@prefix ns1: <http://aimicroservice.derby.ac.uk/> .
ns1:genQualityControl a ns1:Bioinfomatic_genQualityControl ;
    ns1:category ns1:Bioinformatics ;
    ns1:contributor <https://www.derby.ac.uk/staff/hongqing-yu/> ;
    ns1:dependency "matplotlib"@en,
        "pandas"@en,
        "scanpy"@en,
        "seaborn"@en ;
    ns1:description "https://scanpy.readthedocs.io/en/stable/"@en ;
    ns1:formate "py"@en ;
    ns1:framwork "annData_qualityControl"@en ;
    ns1:input [ ns1:paramter [ ns1:iocategory ns1:brain_raw ;
                    ns1:iodatatype ns1:h5ad ;
                    ns1:pid "0"@en ] ] ;
    ns1:licence <https://en.wikipedia.org/wiki/Free-software_license> ;
    ns1:output [ ns1:paramter [ ns1:iocategory ns1:brain_qc ;
                    ns1:iodatatype ns1:h5ad ;
                    ns1:pid "0"@en ] ] ;
    ns1:uri ns1:genQualityControl .
```

Figure 9. Quality control microservice semantic description.

With all the microservices registered, researchers can start expressing the analysis task to stop, interact, and provide feedback at any stage during the process of automatically creating the solution. The researchers can also see visualizations of outputs produced by different steps. Therefore, researchers can provide preferences for selecting microservices if there are options.

A realistic example is that a researcher can specify a clustering task applied to the mouse brain single-cell RNASeq dataset. The framework will first try to see if a single microservice can complete this task. The answer is 'no', because no semantic-matched microservice can take the RNASeq CSV input and provide the clustering output. At this juncture, the microservice that can take the RNASeq CSV will be invoked to process the data into the next step with the output of AnnData. If there are multiple choices in the composition sequence, all possibilities will be invoked to run, unless the previous knowledge in the policies has a priority. The possibilities have multiple solutions at the end for researchers to analyze in order to give professional feedback to the system. The feedback will help greatly with the knowledge graph policies. For example, suppose the researcher gives feedback to the system that UMAP is the better embedding method than t-SNE but has no priority on the clustering methods. In that case, the framework will produce two possible clustering results, shown in Figure 10.

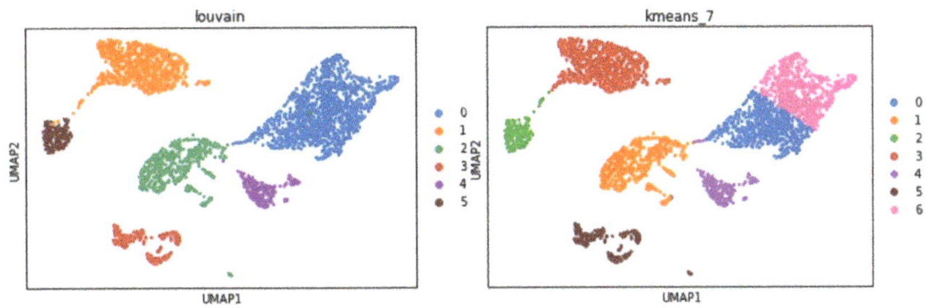

Figure 10. Two clustering outcomes from automatic processes.

7. Discussion

By evaluating the performance of the test scenarios, we believe the combination of KRR and automation of AIMSs offers a viable strategy for developing human-level AI systems. We created an environment with AIMSs capable of handling text, CSV files, and images as default settings. This environment supports data splitting, classification, prediction, and optimization AIMSs. These findings suggest the system's capability to generate and optimize solutions for various tasks by applying or creating knowledge. However, there remain some challenges that future work needs to address:

- The advantage of using a triple KG structure to encode KRR elements lies in its unification, standardization, and adaptability across diverse applications. Nonetheless, as the KG expands, its referencing efficiency diminishes, particularly with intricate graph queries. This inefficiency is exacerbated when different knowledge types are stored separately, making union queries on the graph resource-intensive. A proposed solution is to embed the knowledge graph into a more efficient vector space [44]. To achieve this, we plan on investigating state-of-the-art embedding techniques, such as graph neural networks, that can maintain the relationships between entities while offering efficient querying.
- The current system architecture does not support multimodal inputs pertaining to a singular task (multimodal machine learning). While humans can seamlessly integrate visual, auditory, and other sensory data to accomplish tasks, machines struggle to synthesize multiple data types [45–47]. Moving forward, we aim to explore fusion techniques, both at the feature and decision levels, to facilitate more comprehensive input processing.
- During the initial stages of our manuscript's preparation, Google released research papers detailing the mutation of neural networks (NNs) to handle diverse image classification tasks [16,17]. These papers have illuminated the potential of not just mutating data or services but also the possibility of adding or removing NN hidden layers as a form of knowledge storage for future considerations. Our intent is to delve deeper into the dynamics of such mutations and explore frameworks that allow for flexible and dynamic architectural changes in neural networks.

8. Conclusions

Our proposed Automatic Semantic Machine Learning Microservice (AIMS) framework presents a novel approach to managing the complex demands of machine learning in biomedical and bioengineering research. The AIMS framework utilizes a self-supervised knowledge learning strategy to ensure automatic and dynamic adaptation of machine learning models, making it possible to keep pace with the evolving nature of biomedical research. By placing emphasis on model interpretability and the integration of domain knowledge, the framework facilitates an improved understanding of the decision-making process, enhancing the relevance and applicability of the generated models. A significant finding of this research is our demonstration that knowledge-based systems can play a pivotal role in self-learning AI systems for biomedical research. Such systems offer the capability to store domain-specific knowledge with reusability and bolster the reinforcement learning processes for machines. Furthermore, the potential of these systems extends beyond biomedical research, suggesting applicability to AI applications in other domains.

The three case studies presented underscore the framework's effectiveness in various biomedical research scenarios, demonstrating its capacity to handle different types of data and research questions. As such, the AIMS framework not only offers a robust solution to current challenges in biomedical and bioengineering research but also sets a promising direction for future developments in automated, domain-specific machine learning. Further studies are required to evaluate the AIMS framework's performance across a wider range of biomedical and bioengineering applications and to refine its capabilities for even more efficient and precise knowledge discovery.

Author Contributions: Conceptualization, H.Q.Y.; methodology, H.Q.Y. and S.O.; development and enhancement, H.Q.Y. and S.O.; validation: A.K. and H.Q.Y.; formal analysis, S.O., A.K. and H.Q.Y.; data curation, H.Q.Y.; writing original draft preparation, H.Q.Y.; writing—review and editing, S.O. and A.K.; visualization, H.Q.Y. and S.O. All authors have read and agreed to the published version of the manuscript.

Funding: This research received no external funding.

Data Availability Statement: The full implementation and publicly available data used in this project can be found in GitHub repository (https://github.com/semanticmachinelearning/AISMK, accessed on 20 August 2023).

Conflicts of Interest: The authors declare no conflict of interest.

References

1. Obermeyer, Z.; Emanuel, E.J. Predicting the future—Big data, machine learning, and clinical medicine. *N. Engl. J. Med.* **2016**, *375*, 1216–1219. [CrossRef] [PubMed]
2. Miotto, R.; Wang, F.; Wang, S.; Jiang, X.; Dudley, J.T. Deep learning for healthcare: Review, opportunities and challenges. *Briefings Bioinform.* **2017**, *19*, 1236–1246. [CrossRef] [PubMed]
3. Waring, J.; Lindvall, C.; Umeton, R. Automated machine learning: Review of the state-of-the-art and opportunities for healthcare. *Artif. Intell. Med.* **2020**, *104*, 101822. [CrossRef]
4. Holzinger, A.; Langs, G.; Denk, H.; Zatloukal, K.; Müller, H. Causability and explainability of artificial intelligence in medicine. *Wiley Interdiscip. Rev. Data Min. Knowl. Discov.* **2019**, *9*, e1312. [CrossRef] [PubMed]
5. Zheng, W.; Lin, H.; Liu, X.; Xu, B. A document level neural model integrated domain knowledge for chemical-induced disease relations. *BMC Bioinform.* **2018**, *19*, 328. [CrossRef] [PubMed]
6. Abadi, M.; Agarwal, A.; Barham, P.; Brevdo, E.; Chen, Z.; Citro, C.; Corrado, G.S.; Davis, A.; Dean, J.; Devin, M.; et al. TensorFlow: Large-Scale Machine Learning on Heterogeneous Systems. 2015. Available online: tensorflow.org (accessed on 7 May 2023).
7. Pedregosa, F.; Varoquaux, G.; Gramfort, A.; Michel, V.; Thirion, B.; Grisel, O.; Blondel, M.; Prettenhofer, P.; Weiss, R.; Dubourg, V.; et al. Scikit-learn: Machine Learning in Python. *J. Mach. Learn. Res.* **2011**, *12*, 2825–2830.
8. Google LLC. Google Cloud AutoML. 2021. Available online: https://cloud.google.com/automl/docs (accessed on 24 September 2023).
9. Le, T.T.; Fu, W.; Moore, J.H. Scaling tree-based automated machine learning to biomedical big data with a feature set selector. *Bioinformatics* **2020**, *36*, 250–256. [CrossRef]
10. H2O.ai. H2O AutoML. 2023. Available online: https://docs.h2o.ai/h2o/latest-stable/h2o-docs/automl.html (accessed on 24 September 2023).
11. LeDell, E.; Poirier, S. H2O AutoML: Scalable Automatic Machine Learning. In Proceedings of the 7th ICML Workshop on Automated Machine Learning (AutoML), Vienna, Austria, 12–8 July 2020.
12. Ramsundar, B.; Eastman, P.; Walters, P.; Pande, V. *Deep Learning for the Life Sciences: Applying Deep Learning to Genomics, Microscopy, Drug Discovery, and More*; O'Reilly Media: Sebastopol, CA, USA, 2019.
13. He, J.; Baxter, S.L.; Xu, J.; Xu, J.; Zhou, X.; Zhang, K. The practical implementation of artificial intelligence technologies in medicine. *Nat. Med.* **2019**, *25*, 30–36. [CrossRef]
14. Mustafa, A.; Rahimi Azghadi, M. Automated Machine Learning for Healthcare and Clinical Notes Analysis. *Computers* **2021**, *10*, 24. [CrossRef]
15. Ntoutsi, E.; Fafalios, P.; Gadiraju, U.; Iosifidis, V.; Nejdl, W.; Vidal, M.E.; Ruggieri, S.; Turini, F.; Papadopoulos, S.; Krasanakis, E.; et al. Bias in data-driven artificial intelligence systems—An introductory survey. *WIREs Data Min. Knowl. Discov.* **2020**, *10*, e1356. [CrossRef]
16. Gesmundo, A.; Dean, J. An Evolutionary Approach to Dynamic Introduction of Tasks in Large-scale Multitask Learning Systems. *arXiv* **2022**, arXiv:2205.12755.
17. Gesmundo, A.; Dean, J. muNet: Evolving Pretrained Deep Neural Networks into Scalable Auto-tuning Multitask Systems. *arXiv* **2022**, arXiv:2205.10937.
18. LeCun, Y. A Path Towards Autonomous Machine Intelligence. Open Review. 27 June 2022. Available online: https://openreview.net/pdf?id=BZ5a1r-kVsf (accessed on 24 September 2023).
19. Yao, Q.; Wang, M.; Escalante, H.J.; Guyon, I.; Hu, Y.; Li, Y.; Tu, W.; Yang, Q.; Yu, Y. Taking Human out of Learning Applications: A Survey on Automated Machine Learning. *arXiv* **2018**, arXiv:1810.13306.
20. Jin, H.; Song, Q.; Hu, X. Auto-Keras: An Efficient Neural Architecture Search System. *arXiv* **2018**, arXiv:1806.10282.
21. Sharma, L.; Garg, P.K. Knowledge representation in artificial intelligence: An overview. In *Artificial Intelligence*; CRC: Boca Raton, FL, USA, 2021; pp. 19–28.
22. Cozman, F.G.; Munhoz, H.N. Some thoughts on knowledge-enhanced machine learning. *Int. J. Approx. Reason.* **2021**, *136*, 308–324. [CrossRef]
23. Hu, Z.; Yang, Z.; Salakhutdinov, R.; Xing, E. Deep neural networks with massive learned knowledge. In Proceedings of the 2016 Conference on Empirical Methods in Natural Language Processing, Austin, TX, USA, 1–5 November 2016; pp. 1670–1679.
24. Chen, X.; Jia, S.; Xiang, Y. A review: Knowledge reasoning over knowledge graph. *Expert Syst. Appl.* **2020**, *141*, 112948. [CrossRef]
25. Berners-Lee, T.; Hendler, J.; Lassila, O. The semantic web. *Sci. Am.* **2001**, *284*, 34–43. [CrossRef]
26. Baader, F.; Horrocks, I.; Lutz, C.; Sattler, U. *Introduction to Description Logic*; Cambridge University Press: Cambridge, UK, 2017.

27. Zhang, Q. Dynamic Uncertain Causality Graph for Knowledge Representation and Probabilistic Reasoning: Directed Cyclic Graph and Joint Probability Distribution. *IEEE Trans. Neural Netw. Learn. Syst.* **2015**, *26*, 1503–1517. [CrossRef]
28. Botha, L.; Meyer, T.; Peñaloza, R. The Probabilistic Description Logic. *Theory Pract. Log. Program.* **2021**, *21*, 404–427. [CrossRef]
29. Yu, H.Q.; Reiff-Marganiec, S. Learning Disease Causality Knowledge From the Web of Health Data. *Int. J. Semant. Web Inf. Syst. (IJSWIS)* **2022**, *18*, 1–19. [CrossRef]
30. Zhao, X.; Liu, E.; Yu, H.Q.; Clapworthy, G.J. A Linear Logic Approach to the Composition of RESTful Web Services. *Int. J. Web Eng. Technol.* **2015**, *10*, 245–271. [CrossRef]
31. Allameh Amiri, M.; Serajzadeh, H. QoS aware web service composition based on genetic algorithm. In Proceedings of the 2010 5th International Symposium on Telecommunications, Kauai, HI, USA, 5–8 January 2010; pp. 502–507. [CrossRef]
32. Qiang, B.; Liu, Z.; Wang, Y.; Xie, W.; Xina, S.; Zhao, Z. Service composition based on improved genetic algorithm and analytical hierarchy process. *Int. J. Robot. Autom.* **2018**. [CrossRef]
33. Yu, H.Q.; Zhao, X.; Reiff-Marganiec, S.; Domingue, J. Linked Context: A Linked Data Approach to Personalised Service Provisioning. In Proceedings of the 2012 IEEE 19th International Conference on Web Services, Honolulu, HI, USA, 24–29 June 2012; pp. 376–383. [CrossRef]
34. Dong, H.; Hussain, F.; Chang, E. Semantic Web Service matchmakers: State of the art and challenges. *Concurr. Comput. Pract. Exp.* **2013**, *25*, 961–988. [CrossRef]
35. Publio, G.C.; Esteves, D.; Lawrynowicz, A.; Panov, P.; Soldatova, L.N.; Soru, T.; Vanschoren, J.; Zafar, H. ML-Schema: Exposing the Semantics of Machine Learning with Schemas and Ontologies. *arXiv* **2018**, arXiv:1807.05351.
36. Braga, J.; Dias, J.; Regateiro, F. A machine learning ontology. *Frenxiv Pap.* **2020**, *preprint*. [CrossRef]
37. Tan, C.; Sun, F.; Kong, T.; Zhang, W.; Yang, C.; Liu, C. A Survey on Deep Transfer Learning. In *Proceedings of the Artificial Neural Networks and Machine Learning–ICANN 2018: 27th International Conference on Artificial Neural Networks, Rhodes, Greece, 4–7 October 2018;* Proceedings, Part III 27; Kůrková, V., Manolopoulos, Y., Hammer, B., Iliadis, L., Maglogiannis, I., Eds.; Springer International Publishing: Cham, Switzerland, 2018; pp. 270–279.
38. Filice, R.W.; Kahn, C.E.J. Biomedical Ontologies to Guide AI Development in Radiology. *J. Digit. Imaging* **2021**, *34*, 1331–1341. [CrossRef]
39. Black, S.; Leo, G.; Wang, P.; Leahy, C.; Biderman, S. GPT-Neo: Large Scale Autoregressive Language Modeling with Mesh-Tensorflow (1.0). *Zenodo* **2021**. [CrossRef]
40. Gao, L.; Biderman, S.; Black, S.; Golding, L.; Hoppe, T.; Foster, C.; Phang, J.; He, H.; Thite, A.; Nabeshima, N.; et al. The Pile: An 800GB Dataset of Diverse Text for Language Modeling. *arXiv* **2020**, arXiv:2101.00027.
41. Bhojanapalli, S.; Chakrabarti, A.; Glasner, D.; Li, D.; Unterthiner, T.; Veit, A. Understanding Robustness of Transformers for Image Classification. In Proceedings of the IEEE/CVF International Conference on Computer Vision (ICCV), Montreal, BC, Canada, 11–17 October 2021; pp. 10231–10241.
42. Luecken, M.D.; Theis, F.J. Current best practices in single-cell RNA-seq analysis: A tutorial. *J. Mol. Syst. Biol.* **2019**, *15*, e8746. [CrossRef]
43. Cannoodt, R. Anndata: 'Anndata' for R. 2022. Available online: https://anndata.readthedocs.io/en/latest/ (accessed on 24 September 2023).
44. Le, T.; Huynh, N.; Le, B. Link Prediction on Knowledge Graph by Rotation Embedding on the Hyperplane in the Complex Vector Space. In *Proceedings of the Artificial Neural Networks and Machine Learning–ICANN 2021: 30th International Conference on Artificial Neural Networks, Bratislava, Slovakia, 14–17 September 2021;* Proceedings, Part III 30; Farkaš, I., Masulli, P., Otte, S., Wermter, S., Eds.; Springer International Publishing: Cham, Switzerland, 2021; pp. 164–175.
45. Ramachandram, D.; Taylor, G.W. Deep multimodal learning: A survey on recent advances and trends. *IEEE Signal Process. Mag.* **2017**, *34*, 96–108. [CrossRef]
46. Baltrušaitis, T.; Ahuja, C.; Morency, L.P. Multimodal machine learning: A survey and taxonomy. *IEEE Trans. Pattern Anal. Mach. Intell.* **2018**, *41*, 423–443. [CrossRef] [PubMed]
47. Li, J.; Hong, D.; Gao, L.; Yao, J.; Zheng, K.; Zhang, B.; Chanussot, J. Deep learning in multimodal remote sensing data fusion: A comprehensive review. *arXiv* **2022**, arXiv:2205.01380.

bioengineering

MDPI

Article

Sophisticated Study of Time, Frequency and Statistical Analysis for Gradient-Switching-Induced Potentials during MRI

Karim Bouzrara [1,2], Odette Fokapu [3,4], Ahmed Fakhfakh [1,5] and Faouzi Derbel [6,*]

1 Laboratory of Signals, Systems, Artificial Intelligence and Networks, Technopark of Sfax, Sakiet Ezzit, Sfax 3021, Tunisia; smarts.lab@crns.rnrt.tn (K.B.); contact@enetcom.usf.tn (A.F.)
2 Department of Electrical Engineering, National Engineering School of Sousse, University of Sousse, Erriadh, Sousse 4023, Tunisia
3 UMR CNRS 7338 Biomécanique and Bioingénierie, University of Technology of Compiègne, Dr. Schweitzer Street, 60200 Compiègne, France; odette.fokapu@gmail.com
4 Laboratory of Innovative Technologies, University of Picardie Jules Verne, UR UPJV 3899, Aisne IUT Campus Cuffies-Soissons, 13 Avenue François Mitterrand, 02880 Compiègne, France
5 National School of Electronics and Telecommunications of Sfax, Technopole of Sfax, Sfax 3018, Tunisia
6 Smart Diagnostic and Online Monitoring, Leipzig University of Applied Sciences, Wachterstraße 13, 04107 Leipzig, Germany
* Correspondence: faouzi.derbel@htwk-leipzig.de

Abstract: Magnetic resonance imaging (MRI) is a standard procedure in medical imaging, on a par with echography and tomodensitometry. In contrast to radiological procedures, no harmful radiation is produced. The constant development of magnetic resonance imaging (MRI) techniques has enabled the production of higher resolution images. The switching of magnetic field gradients for MRI imaging generates induced voltages that strongly interfere with the electrophysiological signals (EPs) collected simultaneously. When the bandwidth of the collection amplifiers is higher than 150 Hz, these induced voltages are difficult to eliminate. Understanding the behavior of these artefacts contributes to the development of new digital processing tools for better quality EPs. In this paper, we present a study of induced voltages collected in vitro using a device (350 Hz bandwidth). The experiments were conducted on a 1.5T MRI machine with two MRI sequences (fast spin echo (FSE) and cine gradient echo (CINE)) and three slice orientations. The recorded induced voltages were then segmented into extract patterns called "artefact puffs". Two analysis series, "global" and "local", were then performed. The study found that the temporal and frequency characteristics were specific to the sequences and orientations of the slice and that, despite the pseudo-periodic character of the artefacts, the variabilities within the same recording were significant. These evolutions were confirmed by two stationarity tests: the Kwiatkowski–Phillips–Schmidt–Shin (KPSS) and the time-frequency approach. The induced potentials, all stationary at the global scale, are no longer stationary at the local scale, which is an important issue in the design of optimal filters adapted to reduce MRI artifacts contaminating a large bandwidth, which varies between 0 and 500 Hz.

Keywords: induced potentials; MRI; time and frequency analysis; stationarity test; KPSS test; surrogates; biomedical engineering; image and signal processing; medical image analysis and medical decision-making

Citation: Bouzrara, K.; Fokapu, O.; Fakhfakh, A.; Derbel, F. Sophisticated Study of Time, Frequency and Statistical Analysis for Gradient-Switching-Induced Potentials during MRI. *Bioengineering* **2023**, *10*, 1282. https://doi.org/10.3390/bioengineering10111282

Academic Editors: Hongqing Yu, Alaa AlZoubi, Yifan Zhao and Hongbo Du

Received: 4 September 2023
Revised: 13 October 2023
Accepted: 24 October 2023
Published: 3 November 2023

1. Introduction

Magnetic resonance imaging, like all other imaging techniques, is designed to support research into diseases. A particular feature of MRI is the high contrast of the smooth tissues produced. This means that even the smallest differences in the body's soft tissues (brain, abdominal organs, spinal cord) are clearly visible. Tumors and inflammatory changes, muscles, tendons, intervertebral discs and joints can, therefore, be particularly well-represented. Magnetic resonance imaging (MRI) techniques, like other medical approaches, such as dental implantology [1], the vascularization of engineered tissues [2] and synthetic bone graft substitution (BGS) [3], are constantly evolving with the aim of improving the image quality and broadening the spectrum of applications. These developments have led to considerable improvements in the spatial and temporal resolution of diagnostic MRI images, and to the development of functional and interventional MRIs. Unfortunately, these advances generate sources of artefacts that "pollute" the electrophysiological signals (EPs) acquired simultaneously [4–6]. Indeed, better spatial and temporal resolution requires increasingly strong gradients with shorter and shorter rise times, sometimes associated with higher magnetic fields (3-7T). EPs are essential for patient monitoring and image synchronization in all types of MRI examinations [7,8]. They are also used as a source of additional information in functional MRI. EPs "pollution" is mainly due to the interaction between the electromagnetic devices needed to construct the MRI images and the EPs acquisition devices used for patient monitoring [9,10]. The signal s(t) collected by the electrodes can be modeled as a linear combination of the desired signal EPs, and three main sources of noise [4]: $s(t) = s_{EPs}(t) + n_{RF}(t) + n_{MHD}(t) + n_{GA}(t)$ $s_{EPs}(t)$ is the electrophysiological signal, $n_{RF}(t)$ represents the tensions induced by the RF pulses, $n_M HD(t)$ represents the tensions induced by the RF pulses and $n_{GA}(t)$ are the voltages induced by the gradient switches. The $n_{RF}(t)$ and $n_M HD(t)$ components can be properly reduced by filtering. The voltages induced by gradient switching are difficult to remove because they have large amplitudes and frequencies that fall within the bandwidth of the electrophysiological signals. For several decades the electrocardiogram (ECG) [11] has been used during MRI examinations for patient monitoring and image acquisition synchronization. However, this technique is performed using electronic devices with a low bandwidth (1 Hz–60 Hz) which does not allow provision of a good diagnostic quality ECG. Indeed, a diagnostic ECG requires a wide bandwidth (0.05–150 Hz). Recent studies show the interest in and difficulties associated with this type of collection [12–15]. The surface electromyography signal is another interesting example in the functional MRI of neurological pathologies. Combining the information from the image of a muscle section and the EMG signal collected simultaneously on the surface of the same muscle could provide diagnostic assistance. At present, obtaining a clean EMG signal in the presence of gradients is a major difficulty; its amplitude is low (few µv) and its bandwidth is (0.05–500 Hz). Research in this field confirms the interest in and difficulties of this type of investigation [16–18]. In the cited works, EMG was used only as a control for muscle activity. The authors did not perform a quantitative analysis to correlate the EMG and fMRI signals, due to gradient-induced potentials affecting the EMG signals. The variations in the time-frequency characteristics of the EMG in the presence of gradient artefacts are even more complex. To eliminate the induced potentials that strongly contaminate the broadband electrophysiological signals, it would be interesting to understand the behavior of these artefacts systematically generated by gradient switching. The results of this type of investigation could contribute to the development of new analogue and digital processing tools leading to better quality EPs. The study we propose here falls within this framework and focuses on the characterization of the temporal and frequency variations of the $n_{GA}(t)$ component. The novelty of this work lies in its focus on addressing a specific problem related to the contamination of electrophysiological signals (EPs) during magnetic resonance imaging (MRI) procedures. In this paper, we present a study of induced voltages collected in vitro. In the laboratory, we developed a device that allows collection of only the potentials induced by gradient switching according to different imaging protocols. After analysis of the temporal and frequency properties, we formulated a study of the stationarity of these induced voltages. The fundamental notion that informs the modeling of a temporal process is that of stationarity. Two approaches for testing stationarity were applied to the rewired artefacts: the KPSS stationarity test and the time-frequency stationarity test. The first part of the paper briefly presents the theoretical basis of the two stationarity test methods used. The second part of the paper describes the process of collecting the induced potentials, the method of analysing their variabilities in the time and frequency domains, which is widely used in several fields, like the monitoring and early warning of mine rockbursts [19], and the algorithm for the stationarity study. The results and a discussion of the observations are presented in the last section.

2. State of the Art of Stationaity Test

The stationarity of electrophysiological signals such as EMGs, EEGs and ECGs have often been studied in order to better develop tools for extracting relevant information [20–22]. The stationarity of these same signals recorded in MRI does not seem to have been studied, probably due to the complexity of the different noise sources involved. It seemed appropriate to first study the noise generated by gradient switching using an in vitro approach.

2.1. The Kwiatkowski–Phillips–Schmidt–Shin Stationarity Test

By definition, a signal is considered strictly stationary if, and only if, its statistical moments are independent of time. In practice, it is virtually impossible to verify stationarity in the strict (or strong) sense, mainly because a real physical signal can never be stationary in the strict sense. For this reason, it makes sense to define stationarity in the weak, i.e., second-order, sense. In practice, an acquired signal can be assimilated to a time series whose "trajectory" we observe and analyze in order to qualify it as stationary if it is likely to result from a stationary process. According to the definition of stationarity in the weak sense, non-stationarity can arise from a time dependence of the first-order moment (the expectation) and/or a time dependence of the variance or the auto-covariance. Kwiatkowski et al. proposed hypothesis tests to verify under the null hypothesis that a series is stationary in level η_μ or around a trend η_τ [23]. To this end, a time series is modeled as follows:

$$y_t = \delta t + \xi_t + \varepsilon_t, \tag{1}$$

where ε_t is a stationary error, δt is a deterministic trend and ξ_t is a random walk given by the following equation:

$$\xi_t = \xi_{t-1} + \mu_t \tag{2}$$

where μ_t iid $(0,\sigma_\mu^2)$: under the null hypothesis, the signal is trend-stationary, i.e., $\sigma_\mu^2 = 0$. In the special case where $\delta = 0$, the KPSS test can be used to check that the signal is weak-sense stationary.

2.2. Stationarity Test with a Time-Frequency Approach

The time-frequency approach for testing the stationarity of a time series recommended by Jun et al. [24,25] is very briefly presented below. The starting point of this approach is that second-order stationary processes are a special case of the class of harmonizable processes where time-varying spectra can be defined. When the process under analysis is stationary, its time-varying spectra can be reduced to the classical power spectral density (PSD). This is true for a good choice, such as for the Wigner–Ville Spectrum (WVS). The basic idea underlying the approach used here is, therefore, that, considered over a given period of time, a process is said to be stationary with respect to this scale of observation if its time-varying spectrum does not support any evolution—in other words, if the spectra at all different times are statistically similar to the global spectrum obtained by marginalization. This idea is not new, but the approach advocated is based on the meaning of the difference "local vs. global".

2.2.1. The Time-Frequency Approach

The first element required for the test is a time-frequency representation susceptible to guarantee robust subsequent processing. The choice here will be made on a multi-window spectrogram, which has the advantage of being a good estimator of the theoretical Wigner–

Ville spectrum [26]. Given a signal x(t), the spectrogram is estimated according to [27]:

$$S_{x,K}(t,f) = \frac{1}{k}\sum_{k=1}^{K} S_x^{(h_k)}(t,f) \quad (3)$$

where $\sum_{k=1}^{K} S_x^{(h_k)}$ represents the K spectrograms computed on the signal x(t), taking as short-term windows the successive terms $h_k(t)$ of an (orthonormal) basis of Hermite functions:

$$S_x^{(h_k)}(t,f) = |\int x(s)h_k(s-t)e^{-i2\pi fs}ds|^2 \quad (4)$$

In practice, the average (4) refers to a reduced number of windows, usually between 5 and 10. Another essential element of any window analysis is, of course, the size of the windows, irrespective of their shape. In the present context, the possibility to vary this size provides an intrinsic degree of freedom to the method in order to adjust the horizon of the local analysis with respect to the global time scale set by the total observation time.

2.2.2. Surrogates

The idea of the test is to identify the concept of stationarity with the equivalence of global and local spectral properties. In order to have a quantifiable basis for comparison between the global and local characteristics, the proposed approach is to associate the observed signal with a "stationary" reference in order to be able to reject the stationarity hypothesis, if necessary. To this end, the authors use the interpretation that, for the same spectrum mean, a non-stationary signal differs from a stationary counterpart by a temporal structure whose signature is found in the spectral phase. Thus, given a single observed signal x(t), it is possible to associate a battery of substitutes [28,29] $s_j(t);\ j = 1,\ldots,J$, each having the same power spectrum as the original signal but a stationary time content. It would be enough to replace the original phase of the spectrum by a random phase.

2.2.3. Distances

The idea is to compare the local spectra with the global spectrum. For this purpose, we defined the quantities marginalized in time as follows:

$$\langle S_{y,k}(t_{n,f})\rangle_n = \frac{1}{N}\sum_{n=1}^{N} S_{y,k}(t_{n,f}) \quad (5)$$

Since the signal $y(t) = x(t)$ for J substitutes $y(t) = s_j(t);\ 1,\ldots,J)$, the different time-frequency spectrum was only evaluated at N times t_n, which are a fraction of the equivalent width of the short-term windows. The "distances" J + 1 between the local and the global spectra are derived from this equation:

$$\{C_n^{(y)} = D(S_{y,K}(t_n,.), \langle S_{y,K}(t_n,.)\rangle_n), n = 1,\ldots,N\} \quad (6)$$

where D (.,.) stands for some dissimilarity measure (or "distance") in frequency.

In order to choose a measure of dissimilarity between spectra, the authors adopt a pragmatic attitude which consists in considering the simplest "distances" that have already proved their efficiency in similar contexts. A good choice of measurement is based on two spectra G (f) and H (f),

$$k(G,H) = k_{KL}(\tilde{G},\tilde{H}).(1 + k_{LSD}(G,H)) \quad (7)$$

Combining the Kullback–Leibler divergence

$$k_{KL}(\tilde{G},\tilde{H}) = \int_{\Omega}(\tilde{G(f)} - \tilde{H(f)})log(\frac{\tilde{G}(f)}{\tilde{H}(f)})df \quad (8)$$

Applied to the normalized spectrum $\tilde{G(f)}$ and $\tilde{H(f)}$ from G(f) and H(f) and the log-spectral deviation

$$k_{LSD}(G,H) = \int_{\Omega} |\log \frac{G(f)}{H(f)}| df \tag{9}$$

2.2.4. Stationarity Test

Let us consider $s_j(t), j = 1, \ldots, J$ as the J substitution signals obtained as just described. When they are analyzed as explained above for the original signal x(t), we finally end up with a new collection of distances depending on both the time indices and the randomizations.

$$\{c_n^{s(j)} = k(S_{s_j,K}(t_n,.), \langle S_{s_j,K}(t_n,.)\rangle_n), n = 1, \ldots, J\} \tag{10}$$

To measure the fluctuations in time of the divergences $c_{n_n}^{(.)}$ between local and global spectra, one can use the distance l_2- defined by equation (16):

$$L(g,h) = \frac{1}{N} \sum_{n=1}^{N} (g_n - h_n)^2 \tag{11}$$

For each pair of sequences $\{(g_n h_n); n = 1, \ldots, N\}$. Regarding the intrinsic variability in the proxy data, the dispersion of the divergences under the null hypothesis of stationarity can be measured by the distribution of the J empirical variances

$$\{\theta_0(j) = L(c^{(s_j)}, \langle c^{(s_j)}\rangle_{n=1,\ldots N}), j = 1, \ldots J\} \tag{12}$$

The distribution is used to determine the threshold γ over which the null hypothesis is rejected. The effective test is, therefore, based on the statistics

$$\theta_1 = L(c^{(x)}, \langle c^{(s_j)}\rangle_{n=1,\ldots,N}), j = 1, \ldots, J \tag{13}$$

And takes the form of the unilateral test:

$$\begin{aligned} \theta_1 > \gamma &: \text{"non} - \text{stationarity"}; \\ \theta_1 < \gamma &: \text{"stationarity"} \end{aligned} \tag{14}$$

2.2.5. Index of Non-Stationarity

Test (20) is used to determine the non-stationarity of a signal in terms of its achievement. In the non-stationary case (where the null hypothesis is rejected), it is then possible to define a non-stationary index (INS) according to the following relation:

$$INS = \sqrt{\frac{\theta_1}{\frac{1}{J}\sum_{j=1}^{J} \theta_0(j)}} \tag{15}$$

3. Signal Acquisition and Treatment

3.1. Recording of Induced Potentials

An experimental bench was built in our laboratory to collect the in vitro potentials induced by gradient switching (Figure 1). A detailed description of this bench was published in a previous paper [30]. From an electrophysiological signal generator (A), five signals can be injected simultaneously into the tunnel via the transmitter (B), the optical fiber and the receiver (C). These signals contaminated at the level of the conductive tissue (D), which is placed at the center of the magnet, are detected by the second transmitter (E) and transmitted to the outside of the tunnel via a second optical fiber and the second receiver

(F). The data is stored and processed by station G. The non-MRI-compatible elements (A), (B), (E) and (G) are placed outside the MRI chamber.

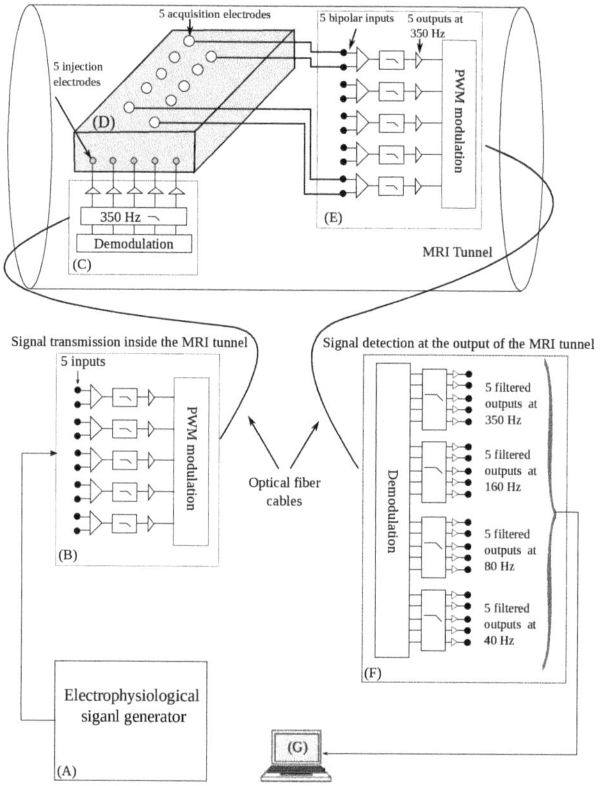

Figure 1. Experimental bench with two "transmitter-receiver" modules: (**B**,**C**) for signal transmission, and (**E**,**F**) for detection. From (**A**), five signals can be injected simultaneously into the tunnel via the transmitter (**B**), the optical fiber and the receiver (**C**). These contaminated signals at the conductive fabric (**D**) are detected by the second transmitter (**E**) and transmitted to the outside of the tunnel via a second optical fiber and the second receiver (**F**). When the generator (**A**) is off and the MRI sequences are activated, the system collects the potentials induced by the gradient switches. The contaminated signals or induced potentials are stored and processed by the station (**G**).

The bandwidth of the set extends from 0.05 Hz to 350 Hz. The bench has 20 channels divided into four frequency bands (40 Hz, 80 Hz, 160 Hz and 350 Hz). It is, thus, possible to analyze the changes in the signal parameters according to the sequences, but also within the different frequency bands. This bench offers different types of experiments. It consists of two MRI-compatible "transmitter-receiver" modules. The first allows EPS signals with known characteristics to be introduced into the MRI tunnel. The signals are injected into a sample of conductive tissue placed in the tunnel. The second module allows the signals to be collected after they have been contaminated by artefacts generated by the imaging sequences owing to the electrodes placed on the conductive tissue. When no signal is injected into the tunnel, the potentials induced after activation of the MRI sequences can be collected. This type of experiment was used in the present work. The study focused on the induced potentials collected at the output of the 350 Hz (broadband) filter, which, therefore, contains a maximum of noise generated by the gradients. The experiments were conducted on a 1.5T MRI system (GE Signa HDxt 1.5T, GE Healthcare) equipped with a 33 mT/m gradient system. To simulate a human body, a conductive tissue model was placed in the MRI tunnel. It was made from salt, gelatine powder and water. By varying the concentrations of salt in the gel, different conductivities of the medium were obtained. For this study, the conductivity was 348 Ω·cm. The induced potentials were acquired with three carbon electrodes (3MTM RedDotTM radiolucent electrode). The induced potentials were sampled at 5 kHz and recorded for a duration of 10 s. The MRI sequences used were FSE and CINE in three slice orientations. Fast spin echo (FSE) (Fov = 30 × 30 cm, TR/TE = 500/12 ms, Matrix = 448 × 512) and cine gradient echo (CINE GE) (Fov = 34 × 25 cm, TR/TE = 9.4/5.1 ms, Matrix = 256 × 128) were used as the MRI sequences.

3.2. Pre-Treatment

Figure 2a shows an example of low-frequency noise recorded without MRI sequence activation. Figure 2b illustrates an example of an induced potential where the amplitude modulation by the noise seen in Figure 2a is observed. There are also bursts of artefacts that appear periodically. These two features were exploited in the studies presented below. The studies focused on a 5-s recording duration for global analysis, and on the artefact puffs for local analysis.

A recording without sequence activation was also performed in order to observe the disturbances caused by the static field B0 and the rest of the measurement environment.

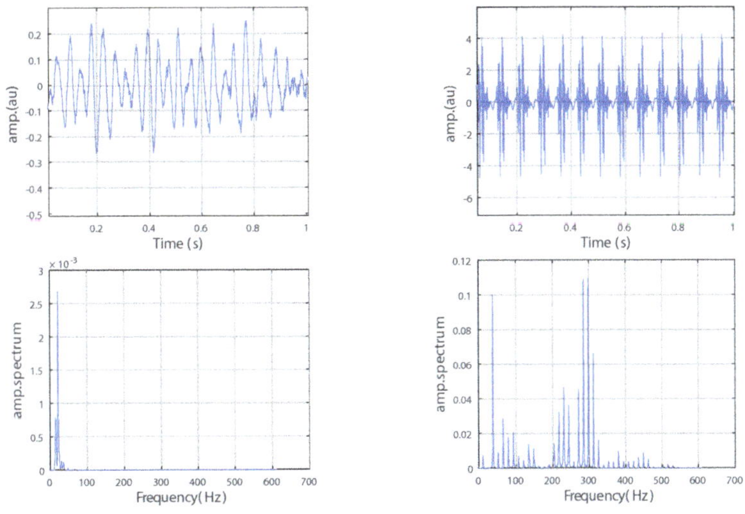

Figure 2. (**a**) Non-sequenced noise and its frequential representation, and (**b**) induced potentials FSE.

3.2.1. Normalization

In order to process the potentials collected in different sequences and slice orientations, the normalization of the data is essential in order to transform the amplitude values of the induced potentials from their original values into comparative scales. In this study the Z-score normalization used was set by the following formula:

$$Y(s) = \frac{X(s) - \mu}{\sigma} \tag{16}$$

Y(s) represents the normalized induced potential, X(s) is the original induced potential μ and σ represents the mean and standard deviation.

3.2.2. Puff Extraction

In order to precisely delineate the artefact bursts, we performed a manual segmentation, which is more accurate for delineating the puffs (Figure 3).

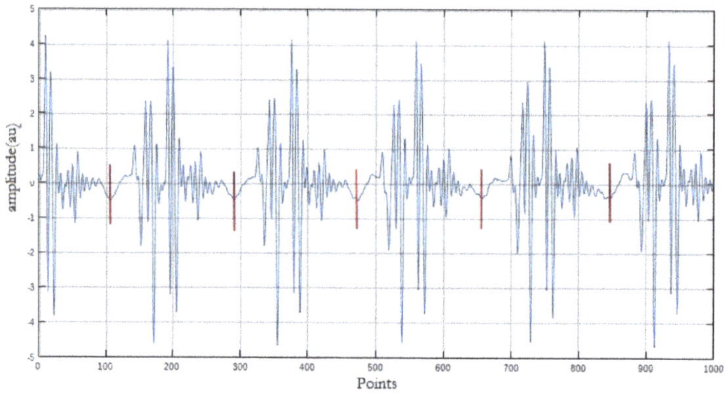

Figure 3. Signal segmentation principle.

The operation was applied to 5 s of induced potential recorded as indicated above for each of the two sequences and according to the three orientations. This gave a total of three "noise signals" per sequence, i.e., six recordings to be segmented. In all, 80 puffs were extracted from each of the six segments, i.e., a total of 480 puffs to be analyzed.

3.2.3. Time and Frequency Analysis of Induced Potentials

The aim here was to verify the expected properties and to highlight the inter-sequence and intra-sequence variabilities. The estimated time and frequency domain characteristics are given below:

- RMS values of the global signal and RMS values of the different bursts.
- Estimation of the average curve of the chirps, and calculation of its RMS value.
- Measurement of the similarity between the puffs by calculating the mean square error between each puff and the mean curve according to the following equation:

$$ern = \frac{1}{\tilde{y}}\sqrt{\sum_{1}^{N}(\hat{y}_i - y_i)^2} \quad (17)$$

This value is normalized to the range $\tilde{y} = y_{max} - y_{min}$.

- The calculation of the power spectral density (PSD) is performed by the Welche–WOSA method, and estimation of the characteristic parameters, the average frequency, the maximum amplitude frequency and the standard deviation of the overall spectrum and on the set of puffs are obtained by segmentation.

3.3. Stationarity Study

3.3.1. Kpss Method

The theoretical approach outlined in Section 2.1 was applied to the different 5 s recordings, which was enough to have sufficient puffs to analyse:

(a) the KPSS of the six 5 s recordings (FSE/axial/ coronal/sagittal-CINE/axial/coronal/sagittal).
(b) the KPSS of the 480 extracted puffs and evaluation of the variabilities by estimating the mean values and standard deviation of the obtained series of values. The values were also grouped and graphed to highlight the degree of stationarity or non-stationarity of the different studied segments of the induced potentials.

3.3.2. Surrogate-Based Method

The algorithm proposed by Bor et al. [24], which is based on the briefly presented approach in Section 2.2, was adapted and applied to our records. The program is implemented in the Matlab language and contains different functions. Below is a presentation of the main steps of the algorithm:

(1) Time-frequency representation: The choice was made for the multi-window Wigner–Ville spectrogram, having successive short-term windows of a Hermite function base. This allows the possibility to adapt the window sizes of our recordings to the MRI sequences.
(2) Surrogate generation: A set of surrogates each having the same power spectral density as the original signal was created. This was achieved by keeping the Fourier transform modulus unchanged but replacing its phase with another randomly taken on $[-\pi, \pi]$.
(3) The stationarity test is based on the distances between the local and global spectra. The distance calculation was carried out by combining the Kullback–Leiber divergence (KL) and log spectral deviation (LSD) methods.

The studied induced potentials show an amplitude modulation by a pseudo sinusoidal low-frequency noise (Figure 2a). Since our induced potentials are similar to the examples of signals used by the authors of the surrogate approach to validate their approach, we were inspired by their approach for the choice of the parameters for the stationarity study. These parameters are as follows:

- Number of substitutes: 5000;
- Number of windows: 5;
- Window size range: [0.03:0.04:0.005:0.07:0.075] adjusted for slice orientation.

The possibility of varying this size provides a degree of freedom intrinsic to the method to allow the local analysis to be adjusted in relation to the global time scale set by the total observation time.

4. Results and Discussion

The temporal and frequency analysis and stationarity studies were applied to a series of induced potential recordings to highlight the variability of the characteristics of the gradient-induced potentials. The different graphs allow a qualitative observation of these variabilities. Tables 1–4 summarize the main parameters calculated for a quantitative analysis of the observed variabilities. The six recordings, according to two sequences and three slice orientations processed throughout this work, are shown in Figure 4. It can be observed that the pseudo-periodicity of each of the six recordings was confirmed by a spectrum of amplitude lines more or less rich according to the imaging sequence. The frequencies of the amplitude maxima are different according to the slice orientations.

4.1. Puff Analysis

The puffs of the potentials obtained by segmentation on the previously described six recordings are shown in Figure 5. The first row shows a 3D representation of the extracted puffs (20 as an example) for each of the FSE and CINE sequences, respectively. The second row shows the respective average puffs. The third row shows the variability in the RMS values of the 80 puffs around the RMS value of the mean puff. The fourth row shows the variability in the average quadratic deviation of each puff from the average puff curve.

The average puffs were estimated for the coronal orientation of the FSE and CINE sequences. The analysis of these puffs shows the variability of the features within a sequence. The RMS values calculated for each puff were compared to the RMS value of the average puff curve. We also compared the root mean square deviation between each puff and the mean puff evaluated from the set of puffs.

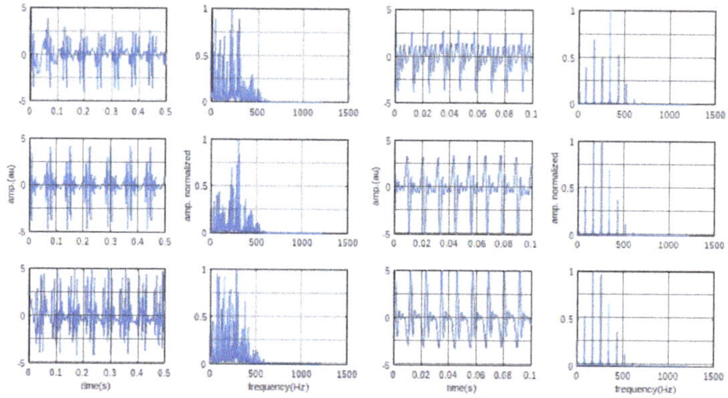

Figure 4. Recorded induced potentials: representation in time and frequency domains. Columns 1 and 2, potentials induced by the FSE sequence (coronal/axial/sagital). Columns 3 and 4, potentials induced by the CINE sequence (coronal/axial/sagittal).

The qualitative analysis of the different graphs in Figure 5 shows that the waveforms of the average puffs from one orientation to another are very different, which is an expected result. The puffs appear in a regular way, but their shape is not identical, as shown in the MSE curves. Table 1 summarizes the values of the calculated RMS and MSE parameters.

The RMS values calculated over the entire duration of the recordings indicate that the orientation of the slices does not determine the level of noise power; it is higher for the axial FSE than the CINE. The dispersion of the RMS values of the puffs is greater for the coronal FSE orientation (0.0644) but the mean value 1.2684 is close to the RMS value calculated on the global recording (1.2566). The variability degree in the morphology of the puffs within the same sequence is more or less significant depending on the sequence and the cutting orientation. The RMS and MSE curves show significant variability for the coronal orientation of the FSE sequence despite the periodic character of the puffs.

For all six recordings, we noted that the estimated average frequency parameters varied significantly from one puff to another for the same slice orientation. This indicates that an in-depth study of the stationarity of the induced potentials is an avenue to explore. The average values of the frequency parameters were calculated in order to compare them with the global values; differences of around 5% were noted.

Table 1. Illustration of RMS and MSE values of FSE and CINE sequences.

		FSE			CINE		
		Axial	**Coronal**	**Sagital**	**Axial**	**Coronal**	**Sagital**
RMS	Global	1.9022	1.2566	1.7004	1.4715	2.2606	2.1468
	[min–max]	[0.8770–1.0017]	[1.0947–1.3463]	[1.2018–1.3360]	[1.3902–1.4610]	[2.1046–2.1855]	[2.0844–2.1394]
	Mean–stdev	0.9393–0.0279	1.2684–0.0644	1.2770–0.0293	1.4154–0.0152	2.1463–0.0181	2.1028–0.0125
MSE	[min–max]	[0.0001–0.0133]	[0.0004–0.0140]	[0.0003–0.0109]	[0.0003–0.0151]	[0.0003–0.0096]	[0.0004–0.0099]
	Mean–stdev	0.0029–0.0036	0.0054–0.0027	0.0036–0.0031	0.0053–0.0037	0.0043–0.0025	0.0043–0.0027

Table 2. Representation of the mean and maximum frequencies and standard deviation for the FSE and CINE puffs.

	FSE			CINE		
Frequency	**Coronal**	**Axial**	**Sagital**	**Coronal**	**Axial**	**Sagital**
F_{mean}	[246.163–254.390]	[87.66–97.451]	[121.364–161.663]	[217.309–231.005]	[254.220–295.720]	[230.406–237.216]
F_{max}	[280.681–287.959]	[9.765–9.765]	[9.548–10.184]	[234.375–234.375]	[234.375–234.375]	[234.375–234.375]
stdev	[0.0013–0.0013]	[0.004–0.005]	[0.0033–0.0049]	[0.0043–0.0051]	[0.0013–0.0015]	[0.0079–0.0084]

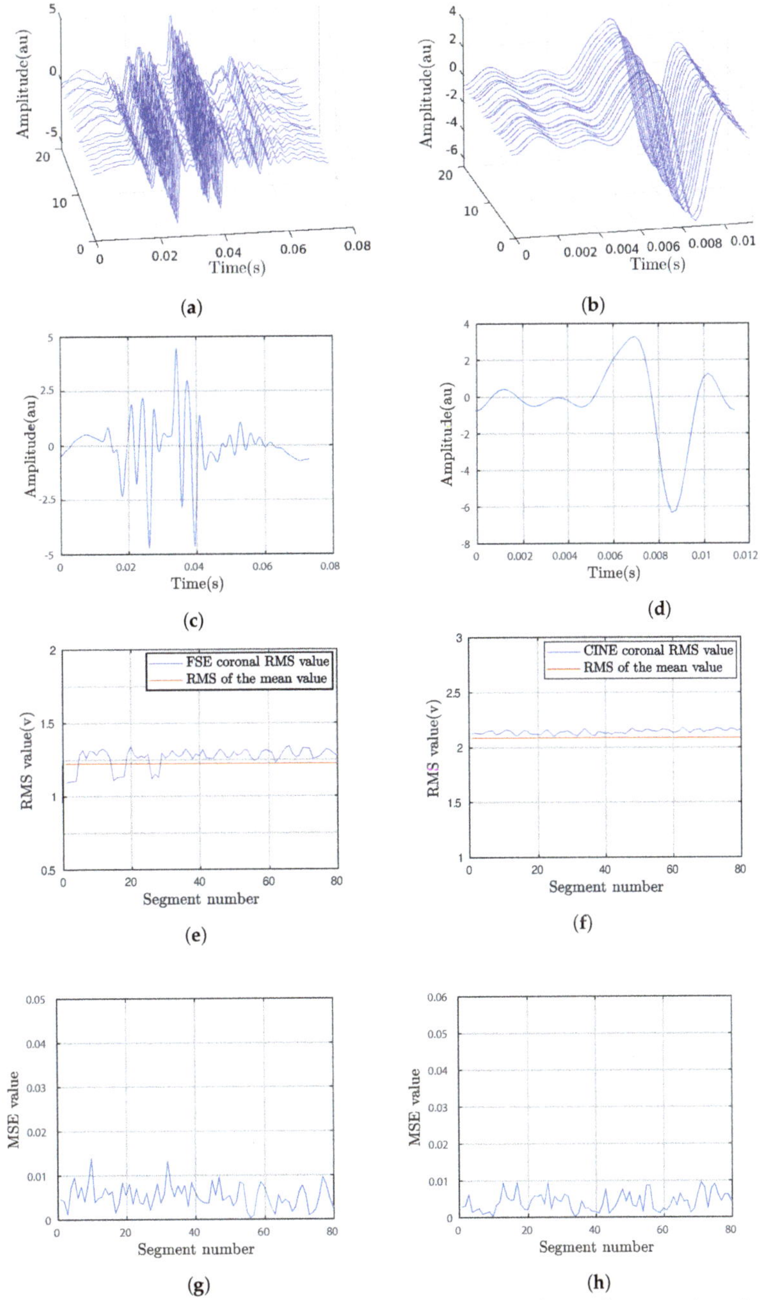

Figure 5. Puffs extracted from the induced potential recordings after activation of the coronal orientation for the FSE and CINE sequences. (**a**,**b**) are the 3D representation of FSE and Cine sequences respectively, (**c**,**d**) are the average puffs of FSE and Cine, (**e**,**f**) the variability in the RMS values around the RMS value of the mean puff and (**g**,**h**) is the variability in the average od quadratic deviation of each puff.

Table 3. Results of the stationarity tests for the two sequences FSE and CINE with the KPSS method and the time-frequency method.

Stationarity Test		FSE			CINE		
		Coronal	Axial	Sagital	Coronal	Axial	Sagital
KPSS test	Statistical value	0.0678	0.0921	0.0403	0.0098	0.0019	0.0020
Surrogates	Theta	0.0089	0.0048	0.0033	0.0072	0.0032	0.0055
	Threshold	0.2698	0.0994	0.4134	0.0311	0.0573	0.0320
	INS	0.0776	0.0941	0.1208	0.0779	0.0377	0.0652
	INS threshold	1.3457	1.3314	1.3341	1.6109	1.5754	1.5632

Table 4. Test results for the FSE puffs.

FSE		KPSS Test		Surrogate Test		
		Statistical Value	Theta	Threshold	INS	INS Threshold
Coronal	Mean–stdev	0.1280–0.1472	0.0051–0.0009	0.0058–0.005	1.2607–0.0984	1.5920–0.0413
Axial	Mean–stdev	0.4376–0.1838	0.0041–0.0004	0.0067–0.0011	1.3963–0.4480	0.8594–0.3205
Sagittal	Mean–stdev	0.3142–0.1475	0.0117–0.0011	0.0067–0.0011	1.9565–0.5376	1.2336–0.2032

4.2. Global and Local Power Spectral Density

The calculated power spectral densities of the induced potentials are shown in Figure 6. They are calculated over the total duration of the recordings (global PSD) and for each series of puffs (local PSD). The first interesting observation on the global PSDs we can make is the very significant variability in the frequency parameters within the same sequence. For example, for the FSE sequence, the average frequency, fmean, is 234.97 Hz, 145.82, and 114.26 for the three orientations, coronal, axial, and sagittal, respectively.

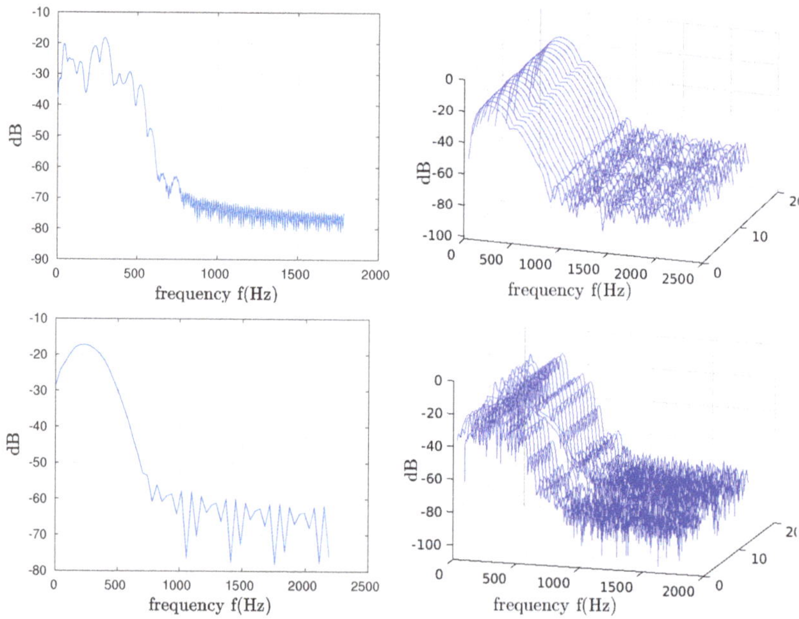

Figure 6. Spectral analysis of the coronal orientation for FSE and CINE sequences. Column 1: global spectrum for each FSE and CINE, respectively. Column 2: spectrum of local segments.

Table 2 shows that the puff-by-puff calculated parameters (local analysis) occupy a quite large interval for each frequency range (min–max). The frequency of the maximum amplitude varies significantly from one puff to the next, whatever the sequence and orientation. This parameter is influenced by the low frequency modulation mentioned in the beginning of the paper. This parameter, whose variations do not depend solely on the morphology of the induced potential, must be analyzed with caution.

4.3. Stationarity

The stationarity study was motivated by the frequency parameters variability observed on the local power spectral densities. The results of this investigation are displayed in this subsection. Two stationarity test processes were proposed: a KPSS test and a time-frequency test. The study was carried out both globally (the total duration of the recording was considered) and locally (each puff extracted from the same recording was analyzed). The local study allows observing the evolution of the stationarity criteria. It was interesting to compare two stationarity-testing methods, one of which uses frequency properties.

For the global study, both methods lead to the results displayed in Table 3, showing the stationarity of the induced potentials whatever the sequence and the cutting orientation. For the KPSS test, the statistical value is effectively below the threshold of 0.1460. The recordings made according to two sequences and three slice orientations, allow highlighting the intra- and inter-sequence variability. The local analysis shows significant levels of variability in the features that cannot be identified in the global analysis. This point was confirmed by both the KPSS and the time-frequency stationarity tests.

For the time-frequency test, a large number of substitutes was taken (JJ = 5000). The stationarity was verified for the six records, and the values of theta and INS were, in fact, lower than the threshold values estimated by the algorithm.

For the local analysis (puff-by-puff evolutionary observations), the stationarity tests results are shown in Table 4. Some puffs are not stationary. It is to be noted that all puffs tested stationary by the time-frequency method were confirmed by the KPSS statistical test. The reverse is not true; in fact, 97.5% of puffs tested stationary by the KPSS method were not confirmed by the time-frequency method. For a given record, the variation range of the stationarity thresholds are generated automatically, while the mean values and the values of the estimated dispersion are shown in Table 4. Depending on the orientation of the cut, the variability in the test parameters can be significant. For example, for the axial orientation, we have a range of [min–max] = [0.1851–0.8744], with a dispersion of 0.18 for the KPSS test; this was confirmed by the threshold values obtained by the time-frequency method. For this slice orientation, no puffs were tested stationary. In contrast, for the coronal orientation, both methods indicate stationarity for 78 of the 80 puffs analyzed.

It should be noted that only the recordings obtained with the FSE sequence are shown because, for the CINE sequence, the time-frequency method did not provide usable results, the duration of the potentials being very short for this type of approach.

For the time-frequency test, the window size is an important parameter in the evaluation of stationarity. The choice of the windows applied was guided by the nature of our "noise signals", which are amplitude modulated, similar to those tested by the authors of this approach [24]. Knowing that for the global study, all our "noise signals" are stationary by the time-frequency approach, we analyzed the influence of the window size for each of the three slice orientations. The evolution of the INS is a function of the number of windows and their sizes. The results are shown in Figure 7.

It was noted that, for the range of variation [0.003–0.05], the obtained curves remain below the INS threshold and show that the induced potentials remain stationary. This approach, which takes into account the frequency characteristics of the signals, is interesting, but the choice of the number of windows and their size is a delicate issue. The number of substitutes is also a parameter that could influence the stationarity results. We tested with 50 and 5000 surrogates and verified the null hypothesis of stationarity. To evaluate the null hypothesis of stationarity, we have taken up the idea advocated in [25], i.e., the representation of the asymptotic histograms of the distributions of θ relative to the surrogates and its fitting by the gamma distribution.

An illustration is given in Figure 8 for two induced potential puffs—one is stationary but the other is not. For the example shown in Figure 8 (5000 and 50 surrogates), the magenta plot is in the middle of the distribution, which proves that the null hypothesis of stationarity is met. The last example shown in Figure 8 is a case of non-stationarity; the statistical value in magenta is far from the distribution.

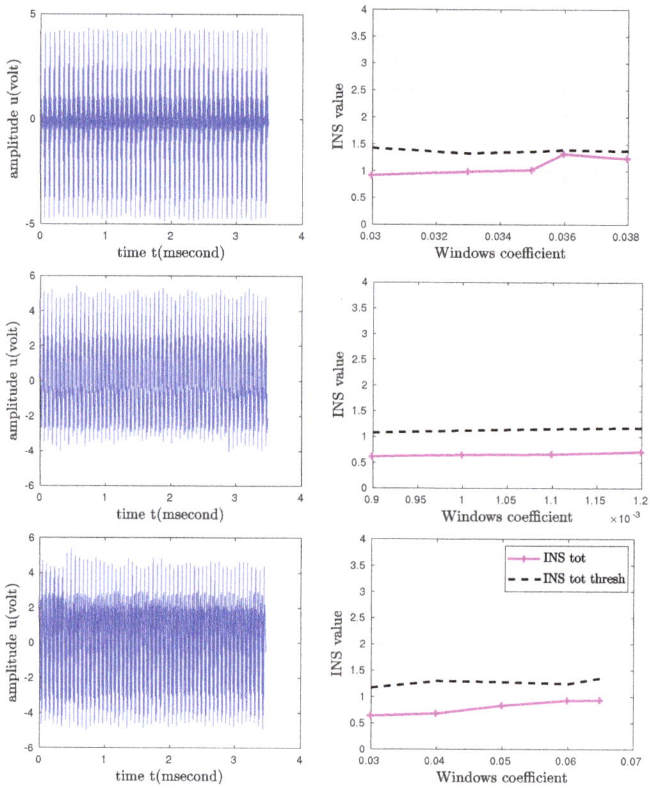

Figure 7. Graphical illustration of INS values in relation to the threshold for the FSE sequence orientations.

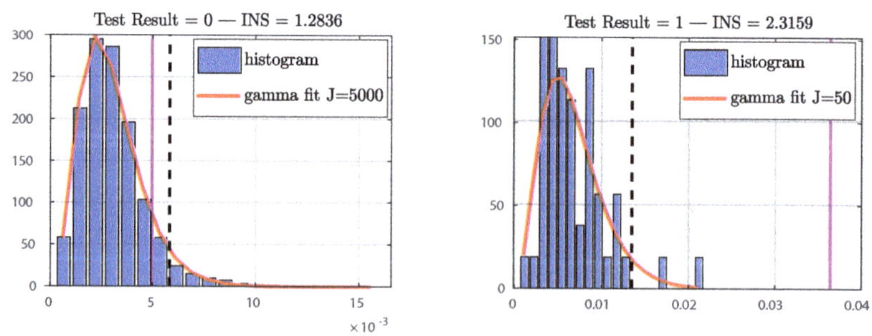

Figure 8. Histogram of Θ(j), surrogate-based distribution and its gamma fitting.

5. Conclusions

These signals are useful to monitor the subject and are also used in the MRI examination process itself, for example, as a source for triggering observation sequences, and in its interpretation by correlation with information obtained in functional MRI. Unfortunately, the technical constraints specific to MRI give rise to sources of artefacts which 'pollute' the electrophysiological signals collected simultaneously. A knowledge of the variability in the characteristics of the artefacts that cause signal disturbances is essential in the choice of strategies to adopt for the development of signal cleaning algorithms. The novelty of this work lies in its systematic investigation of the contamination of EPs by gradient-induced artefacts during MRI scans. It involves the development of a specialized device, detailed analysis of the temporal and frequency properties, and the application of stationarity tests. In this work, an analysis of the induced potentials generated by the gradient switches collected during MRI examination was investigated. The temporal, frequency and statistical characteristics of these artefacts were determined globally and locally. The global analysis was performed on the total recording time and the local analysis on segments extracted from the same recording. The segments designated as puffs of the induced potentials were isolated in accordance with the temporal pseudo-periodicity that characterize these artefacts. It should be noted that the study presented in this paper is just a first step, as the induced studied potentials, collected in vitro, do not have all the characteristics of induced potentials collected in vivo. Forthcoming studies will first focus on the collection of induced potentials in vivo, then on digital filters' modeling. An experimental protocol to isolate segments of the induced potentials generated during the collection of the electrophysiological signals will be developed. The characterizations of the induced potentials is specific to such sequences as MRI sequences. Other MRI sequences, particularly those that generate more artefacts, like true FISP or EP sequences, will be tested in a future study.

Author Contributions: K.B. contributed by identifying state of the art research, methodology, design and implementation, original draft writing, visualization, editing. O.F. contributed by identifying the methodology, validation and writing of sections. A.F. and F.D. way of writing of sections, reviewing, and editing. All authors have read and agreed to the published version of the manuscript.

Funding: The Open Access Publication Funds of HTWK Leipzig.

Institutional Review Board Statement: Not applicable.

Informed Consent Statement: Not applicable.

Data Availability Statement: The original contributions presented in this study are included in this article. Further inquiries can be directed to the corresponding authors.

Conflicts of Interest: The authors declare no conflict of interest.

References

1. Cosoli, G.; Scalise, L.; De Leo, A.; Russo, P.; Tricarico, G.; Tomasini, E.P.; Cerri, G. Development of a Novel Medical Device for Mucositis and Peri-Implantitis Treatment. *Bioengineering* **2020**, *7*, 87. [CrossRef] [PubMed]
2. Hauser, P.V.; Chang, H.M.; Nishikawa, M.; Kimura, H.; Yanagawa, N.; Hamon, M. Bioprinting scaffolds for vascular tissues and tissue vascularization. *Bioengineering* **2021**, *8*, 178. [CrossRef] [PubMed]
3. Mustahsan, V.M.; Anugu, A.; Komatsu, D.E.; Kao, I.; Pentyala, S. Biocompatible Customized 3D Bone Scaffolds Treated with CRFP, an Osteogenic Peptide. *Bioengineering* **2021**, *8*, 199. [CrossRef]
4. Felblinger, J.; Slotboom, J.; Kreis, R.; Jung, B.; Boesch, C. Restoration of electrophysiological signals distorted by inductive effects of magnetic field gradients during MR sequences. *Magn. Reson. Med. Off. J. Int. Soc. Magn. Reson. Med.* **1999**, *41*, 715–721. [CrossRef]
5. Bresch, E.; Nielsen, J.; Nayak, K.; Narayanan, S. Synchronized and noise-robust audio recordings during realtime magnetic resonance imaging scans. *J. Acoust. Soc. Am.* **2006**, *120*, 1791–1794. [CrossRef]
6. Price, D.L.; De Wilde, J.P.; Papadaki, A.M.; Curran, J.S.; Kitney, R.I. Investigation of acoustic noise on 15 MRI scanners from 0.2 T to 3 T. *J. Magn. Reson. Imaging Off. J. Int. Soc. Magn. Reson. Med.* **2001**, *13*, 288–293. [CrossRef]
7. Casale, R.; De Angelis, R.; Coquelet, N.; Mokhtari, A.; Bali, M.A. The Impact of Edema on MRI Radiomics for the Prediction of Lung Metastasis in Soft Tissue Sarcoma. *Diagnostics* **2023**, *13*, 3134. [CrossRef]
8. Feng, M.; Xu, J. Detection of ASD Children through Deep-Learning Application of fMRI. *Children* **2023**, *10*, 1654. [CrossRef]
9. Von Smekal, A.; Seelos, K.; Küper, C.; Reiser, M. Patient monitoring and safety during MRI examinations. *Eur. Radiol.* **1995**, *5*, 302–305. [CrossRef]
10. Shellock, F.G. Patient monitoring in the MRI environment. *Magn. Reson. Proced. Health Eff. Saf.* **2001**, 217–241.

11. Ramírez, W.A.; Gizzi, A.; Sack, K.L.; Filippi, S.; Guccione, J.M.; Hurtado, D.E. On the role of ionic modeling on the signature of cardiac arrhythmias for healthy and diseased hearts. *Mathematics* **2020**, *8*, 2242. [CrossRef]
12. Liu, J.Z.; Zhang, L.; Yao, B.; Yue, G.H. Accessory hardware for neuromuscular measurements during functional MRI experiments. *Magn. Reson. Mater. Physics Biol. Med.* **2001**, *13*, 164–171. [CrossRef] [PubMed]
13. van Duinen, H.; Zijdewind, I.; Hoogduin, H.; Maurits, N. Surface EMG measurements during fMRI at 3T: Accurate EMG recordings after artifact correction. *NeuroImage* **2005**, *27*, 240–246. [CrossRef] [PubMed]
14. Ganesh, G.; Franklin, D.W.; Gassert, R.; Imamizu, H.; Kawato, M. Accurate real-time feedback of surface EMG during fMRI. *J. Neurophysiol.* **2007**, *97*, 912–920. [CrossRef] [PubMed]
15. Van der Meer, J.; Tijssen, M.; Bour, L.; Van Rootselaar, A.; Nederveen, A. Robust EMG–fMRI artifact reduction for motion (FARM). *Clin. Neurophysiol.* **2010**, *121*, 766–776. [CrossRef] [PubMed]
16. Lemieux, L.; Salek-Haddadi, A.; Hoffmann, A.; Gotman, J.; Fish, D.R. EEG-correlated functional MRI: Recent methodologic progress and current issues. *Epilepsia* **2002**, *43*, 64–68. [CrossRef]
17. Allen, P.J.; Josephs, O.; Turner, R. A method for removing imaging artifact from continuous EEG recorded during functional MRI. *Neuroimage* **2000**, *12*, 230–239. [CrossRef]
18. Abächerli, R.; Pasquier, C.; Odille, F.; Kraemer, M.; Schmid, J.J.; Felblinger, J. Suppression of MR gradient artefacts on electrophysiological signals based on an adaptive real-time filter with LMS coefficient updates. *Magn. Reson. Mater. Physics. Biol. Med.* **2005**, *18*, 41–50. [CrossRef]
19. Lou, Q.; Wan, X.; Jia, B.; Song, D.; Qiu, L.; Yin, S. Application Study of Empirical Wavelet Transform in Time–Frequency Analysis of Electromagnetic Radiation Induced by Rock Fracture. *Minerals* **2022**, *12*, 1307. [CrossRef]
20. Xi, Q.; Sahakian, A.V.; Ng, J.; Swiryn, S. Stationarity of surface ECG atrial fibrillatory wave characteristics in the time and frequency domains in clinically stable patients. In Proceedings of the Computers in Cardiology, Thessaloniki, Greece, 21–24 September 2003; pp. 133–136.
21. Lenka, B. Time-frequency analysis of non-stationary electrocardiogram signals using Hilbert-Huang Transform. In Proceedings of the 2015 International Conference on Communications and Signal Processing (ICCSP), Melmaruvathur, India, 2–4 April 2015 ; pp. 1156–1159.
22. Nazmi, N.; Abdul Rahman, M.A.; Yamamoto, S.i.; Ahmad, S.A.; Malarvili, M.; Mazlan, S.A.; Zamzuri, H. Assessment on stationarity of EMG signals with different windows size during isotonic contractions. *Appl. Sci.* **2017**, *7*, 1050. [CrossRef]
23. Kwiatkowski, D.; Phillips, P.C.; Schmidt, P.; Shin, Y. Testing the null hypothesis of stationarity against the alternative of a unit root: How sure are we that economic time series have a unit root? *J. Econom.* **1992**, *54*, 159–178. [CrossRef]
24. Xiao, J.; Borgnat, P.; Flandrin, P. Testing stationarity with time-frequency surrogates. In Proceedings of the 2007 15th European Signal Processing Conference, Poznan, Poland, 3–7 September 2007; pp. 2020–2024.
25. Xiao, J.; Borgnat, P.; Flandrin, P.; Richard, C. Testing stationarity with surrogates-a one-class SVM approach. In Proceedings of the 2007 IEEE/SP 14th Workshop on Statistical Signal Processing, Madison, WI, USA, 26–29 August 2007; pp. 720–724.
26. Flandrin, P. *Time-Frequency/Time-Scale Analysis*; Academic Press: Cambridge, MA, USA, 1998.
27. Bayram, M.; Baraniuk, R.G. Multiple window time-varying spectrum estimation. *Nonlinear Nonstationary Signal Process.* **2000**, *49*, 292–316.
28. Theiler, J.; Eubank, S.; Longtin, A.; Galdrikian, B.; Farmer, J.D. Testing for nonlinearity in time series: The method of surrogate data. *Phys. Nonlinear Phenom.* **1992**, *58*, 77–94. [CrossRef]
29. Keylock, C. Constrained surrogate time series with preservation of the mean and variance structure. *Phys. Rev.* **2006**, *73*, 036707. [CrossRef] [PubMed]
30. Fokapu, O.; El-Tatar, A. An Experimental Setup to Characterize MR Switched Gradient-Induced Potentials. *IEEE Trans. Biomed. Circuits Syst.* **2012**, *7*, 355–362. [CrossRef]

bioengineering MDPI

Article

Comparing the Robustness of ResNet, Swin-Transformer, and MLP-Mixer under Unique Distribution Shifts in Fundus Images

Kazuaki Ishihara and Koutarou Matsumoto *

Biostatistics Center, Kurume University, Kurume 830-0011, Japan; a222ms002i@std.kurume-u.ac.jp
* Correspondence: matsumoto_koutarou@kurume-u.ac.jp; Tel.: +81-942-35-3311

Abstract: Background: Diabetic retinopathy (DR) is the leading cause of visual impairment and blindness. Consequently, numerous deep learning models have been developed for the early detection of DR. Safety-critical applications employed in medical diagnosis must be robust to distribution shifts. Previous studies have focused on model performance under distribution shifts using natural image datasets such as ImageNet, CIFAR-10, and SVHN. However, there is a lack of research specifically investigating the performance using medical image datasets. To address this gap, we investigated trends under distribution shifts using fundus image datasets. Methods: We used the EyePACS dataset for DR diagnosis, introduced noise specific to fundus images, and evaluated the performance of ResNet, Swin-Transformer, and MLP-Mixer models under a distribution shift. The discriminative ability was evaluated using the Area Under the Receiver Operating Characteristic curve (ROC-AUC), while the calibration ability was evaluated using the monotonic sweep calibration error (ECE sweep). Results: Swin-Transformer exhibited a higher ROC-AUC than ResNet under all types of noise and displayed a smaller reduction in the ROC-AUC due to noise. ECE sweep did not show a consistent trend across different model architectures. Conclusions: Swin-Transformer consistently demonstrated superior discrimination compared to ResNet. This trend persisted even under unique distribution shifts in the fundus images.

Keywords: calibration; diabetic retinopathy; distribution shift; fundus image; robustness

Citation: Ishihara, K.; Matsumoto, K. Comparing the Robustness of ResNet, Swin-Transformer, and MLP-Mixer under Unique Distribution Shifts in Fundus Images. *Bioengineering* **2023**, *10*, 1383. https://doi.org/10.3390/bioengineering10121383

Academic Editors: Hongqing Yu, Alaa AlZoubi, Yifan Zhao, Hongbo Du and Larbi Boubchir

Received: 23 October 2023
Revised: 13 November 2023
Accepted: 29 November 2023
Published: 1 December 2023

1. Introduction

1.1. Background

Diabetes is rapidly increasing worldwide, affecting an estimated 537 million people [1]. Approximately 40–45% of people with diabetes are likely to develop diabetic retinopathy (DR) during their lifetime, a leading cause of visual impairment and blindness [2]. It is important to regularly screen patients with diabetes because early symptoms of DR can be subtle and go unnoticed. Early detection of DR can halt its progression; however, manual diagnosis by ophthalmologists is time-consuming and costly. In addition, there is a shortage of ophthalmologists as the number of diabetes cases increases every year, especially in poor regions such as developing countries. To address these issues, automated screening technologies have received considerable attention, and several deep learning models have been developed to detect DR [2–5].

The models used in safety-critical applications, such as medical diagnostic devices, must be both discriminative and well calibrated. A model is well calibrated when its output reflects the true correctness likelihood. Recent studies have shown that modern deep learning models are highly discriminative but poorly calibrated [6,7]. Because safety-critical applications make decisions based on the confidence score of the model, overconfidence and underconfidence are significantly detrimental to patients.

In addition, it is critical that the models used in safety-critical applications are robust to distribution shifts where the distributions of the training and test data differ. Distribution shifts can occur naturally in different real-world settings and are influenced by factors such as different hospitals, cameras, or lighting conditions. Previous studies have shown that although deep learning models are highly accurate when the distributions of the training and test data are the same, they can significantly underperform under distribution shifts [8,9]. Therefore, it is extremely important to evaluate models under distribution shifts assumed to occur in real-world settings.

1.2. Related Works

1.2.1. Discrimination and Calibration Abilities of Deep Learning Models

There have been many reports on the discrimination and calibration capabilities of deep learning models [7,10–14]. Some studies have suggested that modern high-capacity neural networks, such as ResNet, become overconfident by overfitting to a negative log-likelihood (NLL) [7,15]. In contrast, more modern neural networks with non-convolutional architectures, such as the Vision-Transformer (ViT) and MLP-Mixer, have been reported to possess superior discriminative and calibration abilities [10]. Reportedly, the model size and pre-training scale do not fully explain calibration trends and the model architecture is a critical determinant of calibration [10].

1.2.2. Robustness of Deep Learning Models

In recent years, many studies have investigated the robustness of deep learning models, particularly convolutional neural networks (CNNs) and Transformer-based models. While one study suggested that the robustness of CNNs and ViTs is comparable [16], many studies have reported that ViTs are more robust than CNNs [17–20]. One reason for the robustness of ViT is that it has a strong shape bias and is similar to the human cognitive system. Therefore, ViT is expected to have better generalizability than CNNs under distributional shifts [19–21]. The robustness of MLP-Mixers has been inconclusive, with one study suggesting that MLP-Mixers are as robust as CNNs and another suggesting that MLP-Mixers are superior to CNNs [10,22].

1.2.3. Distribution Shift of Fundus Image

In clinical settings, several factors such as lighting conditions, unexpected eye movements, and ocular lesions including cataracts can affect the quality of fundus images, resulting in uneven illumination, blurring, and low contrast. The degradation of fundus images can affect the diagnosis of DR.

Common image corruptions, including Gaussian noise, snow, frost, brightness, and contrast, are often used to induce distribution shifts in natural image datasets [23]. However, there are concerns regarding the application of these image corruptions to fundus images because of the unique noise that occurs in fundus images.

1.3. Objective

Several of the datasets used to investigate model performance under distribution shifts are natural image datasets, such as ImageNet, CIFAR-10, and SVHN, and there is a lack of research investigating model performance under distribution shifts using medical datasets. In this study, we used the retinal fundus image dataset EyePACS [24] to diagnose DR. The purpose of this study was to verify whether the previously reported trends in model performance under distribution shifts remain consistent under unique distribution shifts in fundus images.

2. Materials and Methods

2.1. Dataset

In this study, we used the open-source DR database EyePACS, which contains 35,126 fundus images of both eyes from different racial backgrounds. We obtained permission from the EyePACS office to access and use the dataset for research purposes (Supplementary Materials).

2.2. Outcome

Each image was labeled with DR severity levels based on the International Classification of Diabetic Retinopathy (ICDR) scale. The ICDR scale categorizes DR based on the presence of new blood vessels and distinguishes between non-proliferative diabetic retinopathy (NPDR) and proliferative diabetic retinopathy (PDR). Within NPDR, there are further subcategories: mild, moderate, and severe. Therefore, the ICDR classifies diabetic retinopathy into five levels of severity: no DR, mild NPDR, moderate NPDR, severe NPDR, and PDR. In our study, we adopted a two-class classification task to predict referable DR and defined referable DR as moderate NPDR, severe NPDR, and PDR [25].

2.3. Experimental Pipeline

The experimental pipeline is illustrated in Figure 1. The EyePACS database was randomly divided into training (80%), validation (10%), and test (10%) datasets. The training data were used to fine-tune the pre-trained model, the validation data were used to tune hyperparameters such as the number of epochs, and the test data were used to evaluate model performance on in-distribution data and under distribution shifts. In-distribution refers to scenarios in which the fundus image remains unaltered, whereas a distribution shift refers to scenarios in which noise is introduced into the image. Following previous studies, we induced a distribution shift by introducing three types of noise that can occur in real-world settings during fundus imaging examinations [26]. In previous research, the difference in evaluation metrics before and after the addition of noise has been used as a metric of model robustness [22,23]. Therefore, we adopted the same definition in our study.

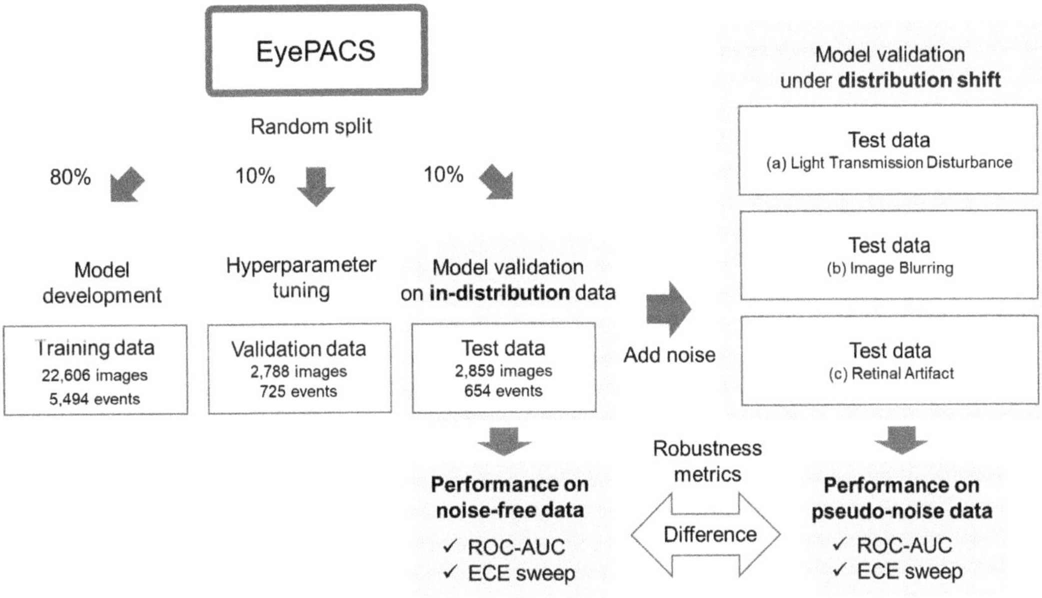

Figure 1. Series of steps from model development to evaluation.

2.4. Preprocessing

The Benjamin Graham method was used to improve the lighting conditions of the fundus image [27,28]. Subsequently, the images were normalized and resized to 224 × 224 pixels. In addition, random horizontal and rotational magnifications were applied.

2.5. Models

To evaluate model robustness, we adopted three model architectures: ResNet, Swin-Transformer, and MLP-Mixer and used pre-trained models (Table 1). The pre-trained models were tuned across all layers. The models were trained for 100 epochs with a batch size of 128. We used 10^{-4} as the base learning rate for the Adam optimizer, along with a default of 20 warm-up iterations and 10^{-5} as the weight decay. During training, the learning rate was reduced by a factor of 10 after 30, 60, and 90 epochs.

Table 1. Pre-trained models used in this study.

Model Name	Model Size (Parameters)
ResNet-50	23.5 M
ResNet-101	42.5 M
ResNet-152	58.1 M
Swin-Transformer (Tiny)	27.5 M
Swin-Transformer (Small)	48.8 M
Swin-Transformer (Base)	86.7 M
Swin-Transformer (Large)	195 M
MLP-Mixer (Base)	59.1 M
MLP-Mixer (Large)	207 M

1. ResNet [29] is a widely used model with a convolutional structure that incorporates residual connections. We used three ResNets with different model sizes: ResNet-50, ResNet-101, and ResNet-152.
2. Swin-Transformer [30] is a model with a non-convolutional structure that implements a hierarchical structure using shifted windows in the Vision-Transformer [31]. We used four Swin-Transformers with different model sizes: Tiny, Small, Base, and Large.
3. MLP-Mixer [32] is a model implemented using only a multilayer perceptron without a convolutional structure or attention mechanism. We used two MLP-Mixers with different model sizes: Base and Large.

2.6. Evaluation

The Area Under the Receiver Operating Characteristic- Curve (ROC-AUC) was used to evaluate the discriminative ability of the models. While the Expected Calibration Error (ECE) is commonly used to evaluate the calibration ability of models [7,33], it has been reported to be an inadequate estimator of calibration error due to its systematic non-negligible bias [34]. Therefore, in this study, we used the monotonic sweep calibration error (ECE sweep) [34], which has been suggested as an estimator with a lower bias than the ECE. The calibration metrics are described in detail below. The robustness of a model was evaluated based on the difference between its performance on noise-free data and that on data with pseudo-noise.

Calibration Metrics

We consider a binary classification with input $X \in \chi$, output $Y = \{0,1\}$, and model $f : X \to [0,1]$ that predicts the confidence score of the true label Y to be 1. Model f is well calibrated if its output correctly reflects the true correctness likelihood. Formally, a perfectly calibrated model satisfies:

$$P(Y = 1|f(X) = p) = p, \forall p \,. \tag{1}$$

True Calibration Error (TCE) is widely used to measure the calibration error by calculating the expected deviation between both sides of Equation (1).

$$\mathrm{TCE}(f) = \mathbb{E}_X[|f(X) - \mathbb{E}_Y[Y|f(X)]|]. \tag{2}$$

$f(X)$ represents the distribution of confidence scores, whereas $\mathbb{E}_Y[Y|f(X)]$ denotes the true calibration curve and illustrates the relationship between the empirical accuracy and confidence scores.

To estimate the TCE of model f, if we are given a finite sample $\{x_i, y_i\}_{i=1}^{n}$, we typically group the sample into equally spaced bins $\{B_m\}_{m=1}^{M}$ based on confidence scores and then calculate the expected difference between the average confidence score $\overline{f_k}$ and the proportion $\overline{y_k}$ where the true label Y is 1.

$$\text{ECE} = \sum_{k=1}^{b} \frac{|B_k|}{n} \left| \overline{f_k} - \overline{y_k} \right| . \tag{3}$$

The calculation of ECE is known to be sensitive to hyperparameters, such as the chosen binning method and the number of bins [35]. In addition, ECE is an inherently biased estimator, and it has been empirically observed that there exists an optimal number of bins that minimizes estimation bias, which tends to increase with the sample size [34]. To address this and determine the optimal number of bins, an ECE sweep is proposed, assuming a monotonically increasing behavior in the true calibration curve and providing a less-biased estimator [34]. The ECE sweep chooses the largest number of bins that preserve monotonicity in the proportion $\overline{y_k}$.

$$\begin{aligned} \text{ECE}_{\text{SWEEP}} &= \sum_{k=1}^{b^*} \frac{|B_k|}{n} \left| \overline{f_k} - \overline{y_k} \right| \text{ where} \\ b^* &= \max\{b | 1 \le b \le n; \forall b' \le b, \overline{y_1} \le \cdots \le \overline{y_{b'}}\}. \end{aligned} \tag{4}$$

2.7. *Distribution Shift of Fundus Image*

Various factors, such as lighting conditions, unexpected eye movements, and ocular pathologies, such as cataracts, can cause uneven illumination, blurring, and low contrast. These elements can significantly degrade the quality of the fundus images. Based on the three realistically occurring factors defined by Shen et al. [26]: (a) Light Transmission Disturbance, (b) Image Blurring, and (c) Retinal Artifact, noise was added to the test data to evaluate the robustness of the model under distribution shifts (Figure 2). To facilitate the interpretation of the effect of noise introduced into retinal images on prediction accuracy, three different noise sources were evaluated one at a time.

2.7.1. Light Transmission Disturbance

The fundus camera was programmed for automatic exposure; however, unstable stray light can cause under/over exposure. Differences in the distance between the fundus and ophthalmoscope can cause uneven illumination due to differences in the sensitivity of certain regions of the image plane. To model these factors, the light transmission disturbance is defined for a clean image $\mathbf{x}$ and its degraded image $\mathbf{x}'$ as

$$\mathbf{x}' = \text{clip}(\alpha(\mathbf{J} \cdot G_L(r_L, \sigma_L) + \mathbf{x}) + \beta; s),$$

where α, β, and s refer to the factors for contrast, brightness, and saturation, respectively. Clip (β; s) represents a clipping function. G_L represents a Gaussian kernel. $\mathbf{J}$ represents the illumination bias to be over- or under-illuminated in a panel centered at (a, b) with a radius of r_L.

2.7.2. Image Blurring

Blurring can be caused by several factors, such as program settings during the fundus imaging procedure, human error, or the presence of cataracts. To model these factors, Image Blurring is defined for a clean image $\mathbf{x}$ and its degraded image $\mathbf{x}'$ as

$$\mathbf{x}' = \mathbf{x} \cdot G_B(r_B, \sigma_B) + n,$$

where G_B is a Gaussian filter with a radius r_B and spatial constant σ_B, and n denotes the additive random Gaussian noise.

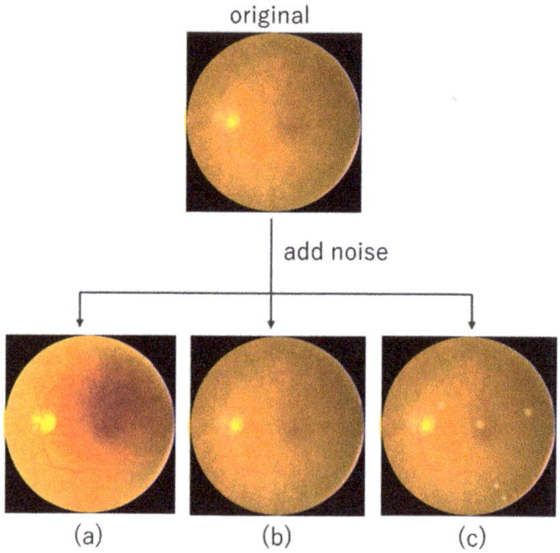

Figure 2. Examples of fundus images without noise and with three types of added noise. (**a**) Light Transmission Disturbance, (**b**) Image Blurring, (**c**) Retinal Artifact.

2.7.3. Retinal Artifact

Imaging in poor conditions can degrade the quality of the fundus image due to dust and grains attached to the lens of the imaging plane. To model these factors, Retinal Artifact is defined for a clean image $\mathbf{x}$ and its degraded image $\mathbf{x}'$ as

$$\mathbf{x}' = \mathbf{x} + \sum_{k}^{K} G_R(r_k/4, \sigma_k) \cdot \mathbf{o}_k,$$

where the Gaussian filter used is G_R, with a specified radius of r_k for object k deemed undesirable and its corresponding variance σ_k. The luminance bias is also represented by $\mathbf{o}_k$.

3. Results

3.1. Model Performance on In-Distribution Data

First, we assessed the discrimination and calibration ability of the models on in-distribution data (Figure 3). The ROC-AUC was the highest for Swin-Transformer (the lowest and highest values for different model sizes: 0.912–0.923), followed by ResNet (0.889–0.904) and MLP-Mixer (0.812–0.831). No significant differences were found between the model architectures in the ECE sweep (Swin-Transformer: 0.012–0.023, ResNet: 0.012–0.034, MLP-Mixer: 0.023–0.026). For all three model architectures, the model size tended to increase with the ROC-AUC value, but this trend was not found in the ECE sweep.

3.2. Model Performance under Distribution Shift

We assessed the discrimination and calibration abilities of the models under three unique distribution shifts in the fundus images (Figure 4). Similar to in-distribution, the ROC-AUC is highest for Swin-Transformer ((a) 0.871–0.887, (b) 0.881–0.918, (c) 0.891–0.911), followed by ResNet ((a) 0.834–0.849, (b) 0.821–0.860, (c) 0.839–0.865) and MLP-Mixer ((a) 0.725–0.753, (b) 0.785–0.812, (c) 0.792–0.820). No consistent trend was found for the ECE sweep as its value for each model differed depending on the distribution shift type and model size (Swin-Transformer: (a) 0.030–0.033, (b) 0.023–0.042, and (c) 0.046–0.060; ResNet: (a) 0.027–0.050, (b) 0.019–0.0430, and (c) 0.043–0.103; MLP-Mixer: (a) 0.060–0.065, (b) 0.036–0.054, and (c) 0.027–0.043).

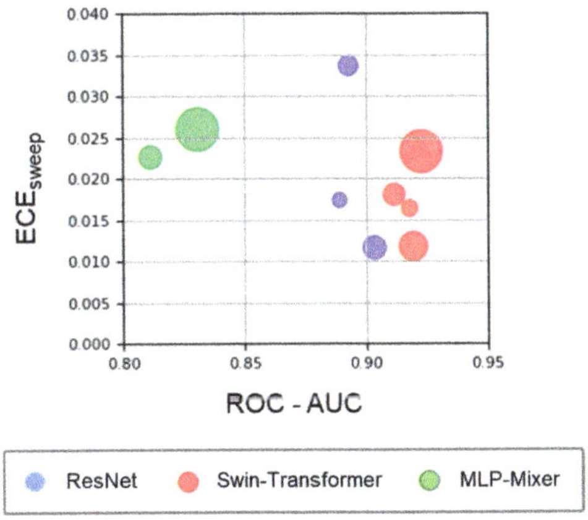

Figure 3. Model performance on in-distribution data. Circle size represents the model size, with the ROC-AUC on the *x*-axis and the ECE sweep on the *y*-axis. The blue, red, and green plots represent the ResNet, Swin-Transformer, and MLP-Mixer architectures, respectively.

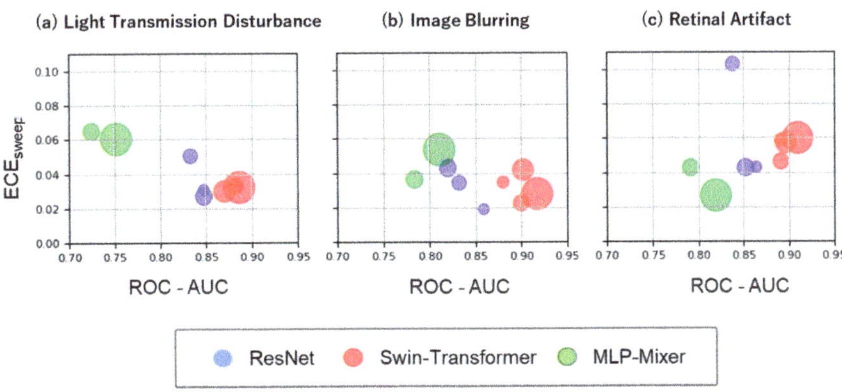

Figure 4. Model performance under three unique distribution shifts in the fundus images: Light Transmission Disturbance (**a**), Image Blurring (**b**), and Retinal Artifact (**c**). Plot details are the same as for Figure 3.

3.3. *Difference in Model Performance between Noise-Free and Pseudo-Noise Data*

We first assessed the robustness of the models in terms of discriminability (Figure 5). The ROC-AUC difference for the distribution shift caused by Light Transmission Disturbance (Figure 5a), compared to in-distribution, was the smallest for Swin-Transformer (0.028–0.049), followed by ResNet (0.041–0.059) and MLP-Mixer (0.078–0.086). The ROC-AUC difference for the distribution shift caused by Image Blurring (Figure 5b) was comparatively small for both the Swin-Transformer (0.005–0.037) and MLP-Mixer (0.019–0.027), followed by ResNet (0.030–0.083). The ROC-AUC difference for the distribution shift caused by Retinal Artifact (Figure 5c) was the lowest for MLP-Mixer (0.011–0.019), followed by Swin-Transformer (0.012–0.027) and ResNet (0.025–0.054). Compared to ResNet, Swin-Transformer showed a smaller reduction in the ROC-AUC across all noise types. We also compared the ROC-AUC reductions across the three distribution shifts within each model; both the Swin-Transformer and MLP-Mixer tended to deteriorate mainly under the distribution shift caused by Light Transmission Disturbance. From the perspective of model size, ResNet tended to increase the reduction in the ROC-AUC with increasing model size. In contrast, Swin-Transformer and MLP-Mixer tended to decrease the reduction in the ROC-AUC with increasing model size.

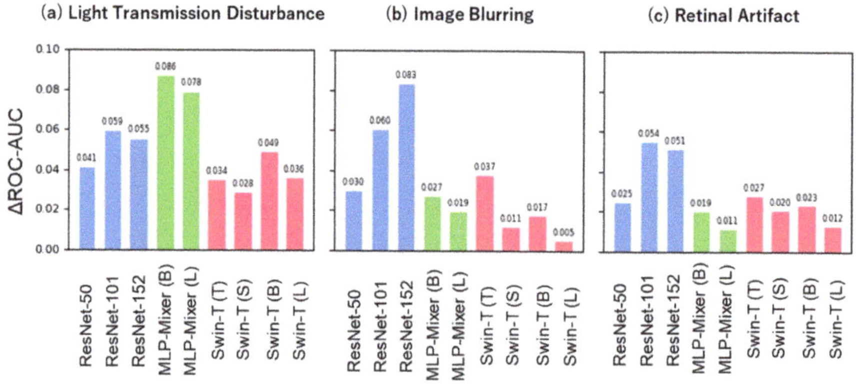

Figure 5. Robustness for discriminative ability under three unique distribution shifts. The *x*-axis showcases each model architecture with different model sizes. The blue, green, and red bars represent the ResNet, MLP-Mixer, and Swin-Transformer architectures, respectively. The *y*-axis highlights the difference in ROC-AUC between distribution shift and in-distribution.

Next, we assessed the robustness of the models in terms of their calibration ability (Figure 6). The difference in the ECE sweep values between distribution shift and in-distribution did not show a consistent trend as the ECE sweep value for each model varied depending on the type of distribution shift and model size (Swin-Transformer: (a) 0.009–0.018, (b) 0.005–0.030, and (c) 0.028–0.045; ResNet: (a) 0.014–0.017, (b) 0.001–0.031, and (c) 0.026–0.069; MLP-Mixer: (a) 0.034–0.042, (b) 0.014–0.028, and (c) 0.001–0.020). In contrast, when comparing the reduction in the ECE sweep across the three distribution shifts within each model, the MLP-Mixer tended to degrade under the distribution shift caused by Light Transmission Disturbance, whereas both ResNet and Swin-Transformer tended to degrade under the distribution shift caused by Retinal Artifact.

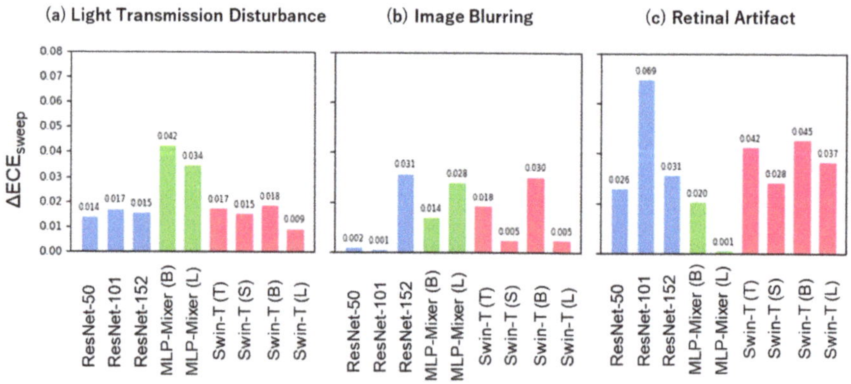

Figure 6. Robustness for calibration ability under three unique distribution shifts. The *x*-axis depicts each model architecture with different model size. The blue, green, and red bars represent the ResNet, MLP-Mixer, and Swin-Transformer architectures, respectively. The *y*-axis represents the difference in ECE sweep between distribution shift and in-distribution.

4. Discussion

The main findings of this study are as follows: Swin-Transformer displayed a consistently higher discriminative ability than ResNet. This trend persisted even under unique distribution shifts in the fundus images. No significant differences were found in the calibration ability between the model architectures and model sizes.

4.1. Model Discrimination and Calibration Ability

Swin-Transformer demonstrated superior discriminative ability under both in-distribution and distribution shifts, followed by ResNet and MLP-Mixer. These results are consistent with those of a previous study using a natural image dataset [22]. Significant findings in retinal images of diabetic retinopathy include capillary aneurysms, beaded expansion, intraretinal microvascular abnormalities, hard exudates, soft exudates, new vessels, and vitreous hemorrhage. These findings are primarily localized and appear at different scales within retinal images. A Swin-Transformer model builds hierarchical feature maps by gradually merging features from adjacent small patches to create representations for larger patches. This hierarchical structure can capture features at different scales ranging from global image features to finer details. This approach may effectively capture the localized and different-scale features of diabetic retinopathy present in retinal images, potentially leading to its high discriminative performance.

Previous studies suggest that non-convolutional models, such as ViT and MLP-Mixer, have a better calibration ability than CNNs in both in-distribution and distribution shifts [10]. In addition, it has been reported that large deep learning models trained with a large number of parameters using negative log-likelihood exhibit overconfidence [7,15]. However, in this study, no significant differences were observed in calibration performance based on model architecture or size. Several previous studies identified factors that affect calibration, including regularization, model size, insufficient data, and imbalanced data [7,12]. As suggested by previous studies, various factors could complexly influence calibration performance, making it challenging to discern differences due to the architecture or model size; therefore, further research is needed.

4.2. Model Robustness

Previous studies have suggested that Transformer-based models are more robust than CNNs in their discriminative and calibration abilities [10,17–20]. Similarly, our study indicates that Swin-Transformer is more robust in its discriminative ability than ResNet as it consistently achieves a smaller reduction in the ROC-AUC across all distribution shifts considered in this study. Previous studies on the robustness of MLP-Mixer compared to CNNs have provided contradicting results [10,22]. Our study could not demonstrate the robustness of MLP-Mixture.

Herein, we considered three types of noise that can occur in fundus images. We hypothesized that Light Transmission Disturbance would primarily affect the texture of images, Image Blurring would affect their shape, and Retinal Artifact would potentially affect both texture and shape. Previous studies have suggested a strong texture bias in CNNs, whereas Transformer-based models, including ViT, indicate a stronger shape bias [19–21]. Therefore, we postulated that ResNet might be particularly susceptible to distribution shifts induced by Light Transmission Disturbance and Swin-Transformer to those induced by Image Blurring. However, our findings did not corroborate these anticipated tendencies (Figures 5 and 6). The discrepancy between our assumptions and results could be due to the fact that Light Transmission Disturbances strongly affect not only the texture but also the shape. Alternatively, the low intensity of the image blurring noise could have resulted in a minimal effect on the shape. Further research is needed to draw definitive conclusions.

The calibration ability of ResNet and Swin-Transformer significantly worsened under the distribution shifts caused by Retinal Artifact (Figure 6c). Previous research has suggested that adversarial attacks on medical images are easier to conduct than on natural images, indicating a vulnerability in deep neural network models developed for medical images [36]. This is because medical images have complex biological textures, resulting in regions of high gradients that are sensitive to small adversarial perturbations. Therefore, in our study retinal artifacts may have behaved similarly to adversarial perturbations, potentially influencing the calibration performance.

In addition, MLP-Mixer was particularly susceptible to distribution shifts caused by Light Transmission Disturbance (Figures 5a and 6a), suggesting that it may be affected by distribution shifts based on principles different from those of ResNet and Swin-Transformer, which requires further investigation.

4.3. Limitations

This study had several limitations. First, because a single dataset was used, additional verification using different fundus image datasets is required to validate the results of this study across all fundus images. Second, models with lower inductive bias, such as the Swin-Transformer and MLP-Mixer, require large amounts of data to improve accuracy; the dataset used in this study may not be large enough for these models to demonstrate their intrinsic capabilities. To mitigate this problem, we fine-tuned the models that were pre-trained on ImageNet. Finally, the recently developed ConvNeXt, a CNN architecture, was reported to exhibit robustness comparable to that of Transformer-based models [16]. Further research is needed to compare the robustness of convolutional and Transformer-based models.

5. Conclusions

In this study, we assessed the performances of the ResNet, Swin-Transformer, and MLP-Mixer models under unique distribution shifts in fundus images. Swin-Transformer demonstrated superior and robust discriminative ability than ResNet under both in-distribution and distribution shifts. In contrast to the previously reported trends in model calibration, this study did not observe significant differences in calibration ability based on model architecture and model size. This discrepancy can be attributed to the unique characteristics of the medical dataset. These findings highlight the need for additional multifaceted validation processes focused on retinal images and additional verification using different fundus image datasets.

Supplementary Materials: Analysis was conducted using Python (version 3.10.11) and PyTorch (version 1.11.0). Further details regarding these programs are available on GitHub (https://github.com/kazuakiishihara/), accessed on 22 October 2023.

Author Contributions: Conceptualization, K.I. and K.M.; methodology, K.I. and K.M.; formal analysis, K.I.; data curation, K.I.; writing—original draft preparation, K.I.; writing—review and editing, K.M.; supervision, K.M.; funding acquisition, K.M. All authors have read and agreed to the published version of the manuscript.

Funding: This study was supported by a Grant-in-Aid for Scientific Research (KAKENHI; Grant number 22K17336) from the Japanese Ministry of Education.

Institutional Review Board Statement: Not applicable.

Informed Consent Statement: Not applicable.

Data Availability Statement: Restrictions apply to the availability of these data. Data were obtained from Kaggle and are available [https://www.kaggle.com/c/diabetic-retinopathy-detection] (accessed on 22 October 2023) with permission from the EyePACS office.

Acknowledgments: We thank the EyePACS office staff for providing the data. We would also like to thank all the professors and students at the Biostatistics Center who provided invaluable advice and insights for our analysis.

Conflicts of Interest: The authors declare no conflict of interest. The funders had no role in the design of the study; in the collection, analyses, or interpretation of data; in the writing of the manuscript; or in the decision to publish the results.

References

1. International Diabetes Federation. *IDF Diabetes Atlas*, 10th ed.; 2021. Available online: https://www.ncbi.nlm.nih.gov/books/NBK581934/ (accessed on 22 October 2023).
2. Gargeya, R.; Leng, T. Automated Identification of Diabetic Retinopathy Using Deep Learning. *Ophthalmology* **2017**, *124*, 962–969. [CrossRef] [PubMed]
3. Gulshan, V.; Peng, L.; Coram, M.; Stumpe, M.C.; Wu, D.; Narayanaswamy, A.; Venugopalan, S.; Widner, K.; Madams, T.; Cuadros, J.; et al. Development and validation of a deep learning algorithm for detection of diabetic retinopathy in retinal fundus photographs. *JAMA* **2016**, *316*, 2402–2410. [CrossRef] [PubMed]
4. Alyoubi, W.L.; Abulkhair, M.F.; Shalash, W.M. Diabetic retinopathy fundus image classification and lesions localization system using deep learning. *Sensors* **2021**, *21*, 3704. [CrossRef] [PubMed]
5. Pires, R.; Avila, S.; Wainer, J.; Valle, E.; Abramoff, M.D.; Rocha, A. A data-driven approach to referable diabetic retinopathy detection. *Artif. Intell. Med.* **2019**, *96*, 93–106. [CrossRef] [PubMed]
6. Hein, M.; Andriushchenko, M.; Bitterwolf, J. Why ReLU networks yield high-confidence predictions far away from the training data and how to mitigate the problem. In Proceedings of the Computer Vision and Pattern Recognition, Long Beach, CA, USA, 16–20 June 2019; pp. 41–50. [CrossRef]
7. Guo, C.; Pleiss, G.; Sun, Y.; Weinberger, Q.K. On calibration of modern neural networks. In Proceedings of the International Conference on Machine Learning, Sydney, Australia, 6–11 August 2017.
8. Taori, R.; Dave, A.; Shankar, V.; Carlini, N.; Recht, B.; Schmidt, L. Measuring robustness to natural distribution shifts in image classification. In Proceedings of the Neural Information Processing Systems, Virtual-Only, 6–12 December 2020.
9. Recht, B.; Roelofs, R.; Schmidt, L.; Shankar, V.; ImageNet, D. Classifiers generalize to ImageNet? In Proceedings of the International Conference on Machine Learning, Long Beach, CA, USA, 9–15 June 2019.
10. Minderer, M.; Djolonga, J.; Romijnders, R.; Hubis, F.; Zhai, X.; Houlsby, N.; Tran, D.; Lucic, M. Revisiting the calibration of modern neural networks. In Proceedings of the Neural Information Processing Systems, Virtual-Only, 6–14 December 2021.
11. Ovadia, Y.; Fertig, E.; Ren, J.; Nado, Z.; Sculley, D.; Nowozin, S.; Dillon, J.; Lakshminarayanan, B.; Snoek, J. Can you trust your Model's uncertainty? Evaluating predictive uncertainty under dataset shift. In Proceedings of the Neural Information Processing Systems, Vancouver, BC, Canada, 8–14 December 2019.
12. Wang, D.; Feng, L.; Zhang, M.L. Rethinking calibration of deep neural networks: Do not be afraid of overconfidence. In Proceedings of the Neural Information Processing Systems, Virtual-Only, 6–14 December 2021.
13. Krishnan, R.; Tickoo, O. Improving model calibration with accuracy versus uncertainty optimization. In Proceedings of the Neural Information Processing Systems, Virtual-Only, 6–12 December 2020.
14. Karandikar, A.; Cain, N.; Tran, D.; Lakshminarayanan, B.; Shlens, J.; Mozer, M.C.; Roelofs, B. Soft calibration objectives for neural networks. In Proceedings of the Neural Information Processing Systems, Virtual-Only, 6–14 December 2021.
15. Mukhoti, J.; Kulharia, V.; Sanyal, A.; Golodetz, S.; Torr, P.; Dokania, P. Calibrating deep neural networks using focal loss. In Proceedings of the Neural Information Processing Systems, Virtual-Only, 6–12 December 2020.
16. Pinto, F.; Torr, P.; Dokania, P. An impartial take to the CNN vs transformer robustness contest. In Proceedings of the European Conference on Computer Vision, Tel Aviv, Israel, 23–27 October 2022.
17. Paul, S.; Chen, P. Vision transformers are robust learners. In Proceedings of the AAAI Conference on Artificial Intelligence, Vancouver, BC, Canada, 20–27 February 2022; Volume 36, pp. 2071–2081. [CrossRef]
18. Bhojanapalli, S.; Chakrabarti, A.; Glasner, D.; Li, D.; Unterthiner, T.; Veit, A. Understanding robustness of transformers for image classification. In Proceedings of the International Conference on Computer Vision, Virtual-Only, 11–17 October 2021; pp. 10211–10221. [CrossRef]
19. Zhang, C.; Zhang, M.; Zhang, S.; Jin, D.; Zhou, Q.; Cai, Z.; Zhao, H.; Liu, X.; Liu, Z. Delving deep into the generalization of vision transformers under distribution shifts. In Proceedings of the Computer Vision and Pattern Recognition, New Orleans, LA, USA, 19–24 June 2022; pp. 7267–7276. [CrossRef]
20. Naseer, M.; Ranasinghe, K.; Khan, S.; Hayat, M.; Khan, F.; Yang, M. Intriguing properties of vision transformers. In Proceedings of the Neural Information Processing Systems, Virtual-Only, 6–14 December 2021.
21. Geirhos, R.; Rubisch, P.; Michaelis, C.; Bethge, M.; Wichmann, F.; Brendel, W. ImageNet-trained CNNs are biased towards texture; increasing shape bias improves accuracy and robustness. In Proceedings of the International Conference on Learning Representations, New Orleans, LA, USA, 6–9 May 2019.
22. Morrison, K.; Gilby, B.; Lipchak, C.; Mattioli, A.; Kovashka, A. Exploring corruption robustness: Inductive biases in vision transformers and MLP-mixers. In Proceedings of the International Conference on Machine Learning, Virtual-Only, 18–24 July 2021.
23. Hendrycks, D.; Dietterich, T. Benchmarking neural network robustness to common corruptions and perturbations. In Proceedings of the International Conference on Learning Representations, New Orleans, LA, USA, 6–9 May 2019.

24. Cuadros, J.; Bresnick, G. EyePACS: An adaptable telemedicine system for diabetic retinopathy screening. *J. Diabetes Sci. Technol.* **2009**, *3*, 509–516. [CrossRef]
25. Ghanchi, F.; Bailey, C.; Chakravarthy, U.; Cohen, S.; Dodson, P.; Gibson, J.; Menon, G.; Muqit, M.; Pilling, R.; Olson, J.; et al. Diabetic Retinopathy Guidelines. 2012. Available online: https://www.rcophth.ac.uk/wp-content/uploads/2021/08/2012-SCI-267-Diabetic-Retinopathy-Guidelines-December-2012.pdf (accessed on 22 October 2023).
26. Shen, Z.; Fu, H.; Shen, J.; Shao, L. Modeling and enhancing low-quality retinal fundus images. *IEEE Trans. Med. Imaging* **2021**, *40*, 996–1006. [CrossRef] [PubMed]
27. Ratthachat, C. APTOS: Eye Preprocessing in Diabetic Retinopathy (Kaggke Report). 2019. Available online: https://www.kaggle.com/code/ratthachat/aptos-eye-preprocessing-in-diabetic-retinopathy/comments (accessed on 22 October 2023).
28. Graham, B. Kaggle Diabetic Retinopathy Detection Competition Report. 2015. Available online: https://www.kaggle.com/c/diabetic-retinopathy-detection (accessed on 22 October 2023).
29. He, K.; Zhang, X.; Ren, S.; Sun, J. Deep residual learning for image recognition. In Proceedings of the Computer Vision and Pattern Recognition, Las Vegas, NV, USA, 26 June 2016; pp. 770–778. [CrossRef]
30. Liu, Z.; Lin, Y.; Cao, Y.; Hu, H.; Wei, Y.; Zhang, Z.; Lin, S.; Guo, B. Swin transformer: Hierarchical vision transformer using shifted windows. In Proceedings of the International Conference on Computer Vision, Virtual-Only, 11–17 October 2021; pp. 9992–10002. [CrossRef]
31. Dosovitskiy, A.; Beyer, L.; Kolesnikov, A.; Weissenborn, D.; Zhai, X.; Unterthiner, T.; Dehghani, M.; Minderer, M.; Heigold, G.; Gelly, S.; et al. An image is worth 16x16 words: Transformers for image recognition at scale. In Proceedings of the International Conference on Learning Representations, Virtual-Only, 3–7 May 2021.
32. Tolstikhin, I.O.; Houlsby, N.; Kolesnikov, A.; Beyer, L.; Zhai, X.; Unterthiner, T.; Yung, J.; Steiner, A.; Keysers, D.; Uszkoreit, J.; et al. MLP-Mixer: An all-MLP architecture for vision. In Proceedings of the Neural Information Processing Systems, Virtual-Only, 6–14 December 2021.
33. Naeini, M.P.; Cooper, G.; Hauskrecht, M. Obtaining Well Calibrated Probabilities Using Bayesian Binning. In Proceedings of the Twenty-Ninth AAAI Conference on Artificial Intelligence, Austin, TX, USA, 25–30 January 2015; p. 29. [CrossRef]
34. Roelofs, R.; Cain, N.; Shlens, J.; Mozer, M.C. Mitigating bias in calibration error estimation. In Proceedings of the International Conference on Artificial Intelligence and Statistics, Virtual-Only, 28–30 March 2022.
35. Nixon, J.; Dusenberry, M.W.; Zhang, L.; Jerfel, G.; Tran, D. Measuring calibration in deep learning. In Proceedings of the Computer Vision and Pattern Recognition, Las Vegas, NV, USA, 16–20 June 2019.
36. Ma, X.; Niu, Y.; Gu, L.; Wang, Y.; Zhao, Y.; Bailey, J.; Lu, F. Understanding adversarial attacks on deep learning based medical image analysis systems. *Pattern Recognit.* **2021**, *110*, 107332. [CrossRef]

bioengineering

MDPI

Article

ClearF++: Improved Supervised Feature Scoring Using Feature Clustering in Class-Wise Embedding and Reconstruction

Sehee Wang [1], So Yeon Kim [1,2] and Kyung-Ah Sohn [1,2,*]

[1] Department of Artificial Intelligence, Ajou University, Suwon 16499, Republic of Korea; wsh0509@ajou.ac.kr (S.W.); jebi1771@ajou.ac.kr (S.Y.K.)
[2] Department of Software and Computer Engineering, Ajou University, Suwon 16499, Republic of Korea
* Correspondence: kasohn@ajou.ac.kr

Abstract: Feature selection methods are essential for accurate disease classification and identifying informative biomarkers. While information-theoretic methods have been widely used, they often exhibit limitations such as high computational costs. Our previously proposed method, ClearF, addresses these issues by using reconstruction error from low-dimensional embeddings as a proxy for the entropy term in the mutual information. However, ClearF still has limitations, including a nontransparent bottleneck layer selection process, which can result in unstable feature selection. To address these limitations, we propose ClearF++, which simplifies the bottleneck layer selection and incorporates feature-wise clustering to enhance biomarker detection. We compare its performance with other commonly used methods such as MultiSURF and IFS, as well as ClearF, across multiple benchmark datasets. Our results demonstrate that ClearF++ consistently outperforms these methods in terms of prediction accuracy and stability, even with limited samples. We also observe that employing the Deep Embedded Clustering (DEC) algorithm for feature-wise clustering improves performance, indicating its suitability for handling complex data structures with limited samples. ClearF++ offers an improved biomarker prioritization approach with enhanced prediction performance and faster execution. Its stability and effectiveness with limited samples make it particularly valuable for biomedical data analysis.

Keywords: feature selection; feature scoring; information theory; entropy; mutual information (MI); dimension reduction; low-dimensional embedding; reconstruction error; principal component analysis (PCA); clustering

Citation: Wang, S.; Kim, S.Y.; Sohn, K.-A. ClearF++: Improved Supervised Feature Scoring Using Feature Clustering in Class-Wise Embedding and Reconstruction. *Bioengineering* **2023**, *10*, 824. https://doi.org/10.3390/bioengineering10070824

Academic Editor: Luca Mesin

Received: 19 May 2023
Revised: 28 June 2023
Accepted: 4 July 2023
Published: 10 July 2023

1. Introduction

In the field of bioinformatics, accurate disease classification is crucial for effective diagnosis and treatment. Furthermore, the precise identification and selection of relevant biomarkers is essential to predict disease risk or aid in drug development [1]. As a result, a significant amount of research is currently being conducted in biomarker detection. Feature selection methods [2–4] are widely used in this context to identify and prioritize biomarkers from large and complex datasets [5,6]. These methods are particularly valuable in bioinformatics, where datasets often have a high number of features relative to the number of samples. By reducing the dimensionality of such data, feature selection can help identify the most informative biomarkers, facilitating accurate disease classification. Many feature selection algorithms have been developed to precisely select the most relevant biomarkers. This is crucial to better understand the underlying mechanisms of disease development and prognosis and to develop more targeted therapies.

Feature selection methods can be broadly classified into supervised and unsupervised approaches, where supervised approaches utilize class labels to identify relevant features, while unsupervised approaches do not [7]. As supervised approaches are more suitable for targeting specific diseases and finding relevant biomarkers, this study utilizes supervised feature selection methods for effective biomarker prioritization. There are various supervised feature selection methods, such as statistical methods [8,9], similarity-based approaches [10,11], and information-theoretic methods. Information-theoretic methods perform feature selection by quantifying the amount of mutual information, which is a measure of entropy and conditional dependencies between data variables and their labels. Information-theoretic methods have been widely studied for feature selection, and their effectiveness has been demonstrated by promising experimental results [12–16]. Recently, innovative approaches like MI-VIF [17] have emerged, which combine variance inflation factor and mutual information, offering a solution to the collinearity problem that leads to unstable parameter estimation. In addition, a methodology named Relevance based on Weight Feature Selection (RWFS) [18] has been proposed. This method is based on two types of changed ratios in relation to feature relevance evaluation: one for the undetermined amount of information and the other for the established amount of information. These strategies have demonstrated their effectiveness by improving performances

However, these methods often suffer from high computational costs, and they may require discretization of continuous variables, which may lead to information loss [19]. To address these issues, we have previously proposed ClearF [20], which uses the reconstruction error of a low-dimensional embedding method as a proxy for the mutual information. ClearF assigns supervised scores to features by applying unsupervised class-wise low-dimensional embedding, which has been demonstrated to be effective in several benchmark datasets. However, ClearF has a limitation in that the selection process of the bottleneck layer is not transparent, requiring the selection of feature size in advance, followed by a greedy search. Consequently, the process can be complicated, unstable, and time-consuming, depending on the experimental setup. Furthermore, due to the partitioning of the entire dataset based on class labels and the subsequent embedding of each partition, the sample size becomes significantly smaller. This may introduce the risk of generating unstable outcomes during the feature selection process.

In this paper, we propose ClearF++ to address the limitations of ClearF. ClearF++ simplifies the process of determining the number of uncertain bottleneck layers and further improves performance through feature clustering. First, we propose a method to increase convenience and stabilize the process by simply fixing the number of bottleneck layers to a single value. In addition, we apply a feature-wise clustering method to mitigate the problem of embedding too many features at once and only reflecting the importance of a few features. This method allows for the selection of important features by clustering similar features together, thus reducing the number of embedded features. In summary, ClearF++ addresses the limitations of ClearF by simplifying the selection process of bottleneck layers and improving performance through feature clustering. Figure 1 illustrates the proposed architecture, and the entire process is shown in the pseudocode presented in Algorithm 1.

Algorithm 1 Algorithm ClearF++: Supervised feature scoring method using feature clustering in the class-wise embedding and reconstruction method.

function CLEARF++(X, Y, k, l)
  Input:
    $X = \{X_1, X_2, \ldots, X_s\} \in \mathbb{R}^{n \times s}$: Data matrix (n features and s samples)
    $Y = \{y_1, y_2, \ldots, y_s\}$: Label vector
    k: Number of clusters
    l: Number of classes
  Output:
    $F = \{f_1, f_2, \ldots, f_n\}$: Feature scores

  Perform feature-wise clustering on data X:
  Apply DEC clustering method that divides n features into k clusters to obtain $C = \{C_1, C_2, \ldots, C_k\}$, where each cluster $C_i \in \mathbb{R}^{c_i \times s}$.

Algorithm 1 *Cont.*

for $i = 1, \ldots, k$ **do**
$F_i \leftarrow$ CLEARF$(C_i, Y, l, 1)$
end for
Aggregate feature scores for each cluster and rank features to obtain F
return F
end function

function CLEARF(X, Y, l, d)
Input:
$X = \{X_1, X_2, \ldots, X_s\} \in \mathbb{R}^{n \times s}$: Data matrix
$Y = \{y_1, y_2, \ldots, y_s\}$: Label vector
l: Number of classes
d: Number of components
Output:
$F = \{f_1, f_2, \ldots, f_n\}$: Feature scores

Using label vector Y, divide X into $L = \{L_1, L_2, \ldots, L_l\}$, where each divided data $L_j \in \mathbb{R}^{n \times l_j}$.
Perform low-dimensional embedding on X with d components and reconstruct to calculate the feature-wise reconstruction error:
$R_X = \{r_{(X,1)}, r_{(X,2)}, \ldots, r_{(X,n)}\}$
for $j = 1, \ldots, l$ **do**
Perform low-dimensional embedding on L_j and reconstruct to calculate the feature-wise reconstruction error:
$R_j = \{r_{(j,1)}, r_{(j,2)}, \ldots, r_{(j,n)}\}$
end for
$R_{\text{sum}} = \text{sum}(R_1, R_2, \ldots, R_l)$
$F = R_X$-R_{sum}
return F
end function

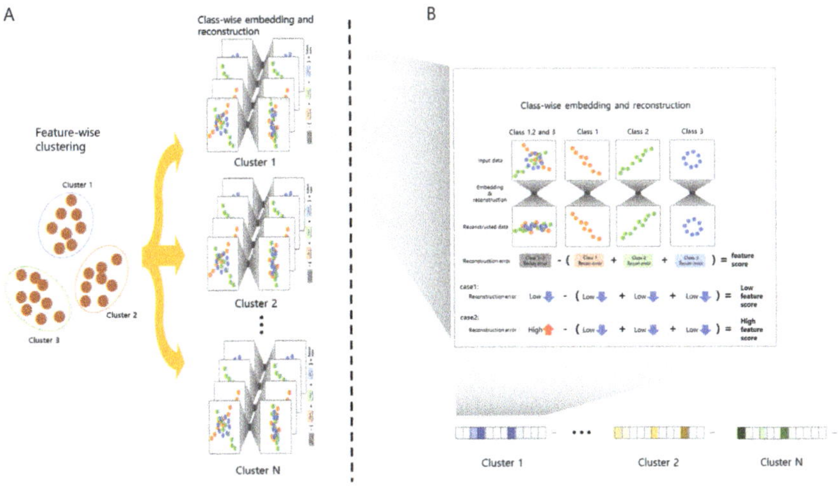

Figure 1. Overview of ClearF++, a supervised feature scoring method that utilizes feature clustering in a class-wise embedding and reconstruction method: (**A**) Description of how the entirety of the data are divided into multiple partitions using feature-wise clustering. (**B**) Description of the process of calculating feature importance using ClearF++.

2. Materials and Methods

2.1. ClearF-One: Simplifying Bottleneck Layer Selection

To tackle the instability issue in the bottleneck layer selection process within ClearF, we propose a refined approach called ClearF-one. In this modified method, the bottleneck layer is set to a single layer rather than employing a greedy search to determine the optimal number of bottleneck layers. As the size of the bottleneck layer increases, a broader range of information is selected, resulting in the dilution of focusing important parts. By constraining the number of layers to one, only the most informative features from each class's embedding are selected, aligning with the theoretical foundation of ClearF. In summary, ClearF-one serves as an enhanced version of ClearF that addresses instability in the feature selection process by simplifying the bottleneck layer to a single layer.

2.2. ClearF++: Advanced Feature Selection via Feature-Wise Clustering

As described above, ClearF-one fixes the bottleneck layer to a single layer, resulting in features with strong signals for each class that are likely to have high scores. However, ClearF-one can be disadvantageous in selecting multiple features due to the limited amount of expressed information. To overcome this limitation, we propose a novel method that divides the data into several partitions through feature-wise clustering and applies ClearF-one to each cluster. As shown in Figure 1A, feature-wise clustering is performed to divide the data into units of each cluster with similar features. When we perform feature-wise clustering, the Deep Embedding Clustering (DEC) method [21] is applied, which is a method of unsupervised learning that combines deep neural networks with clustering algorithms. Next, ClearF-one is applied to each of the clustered data to calculate the feature score. Finally, ClearF++ produces a high feature score when it exhibits a significant difference between classes, such as in case 2 of Figure 1B. Features with no significant difference between classes, such as case 1, are not scored high. This approach allows the most informative features to produce high scores by calculating a class-wise reconstruction error. The above process is performed for each cluster, as depicted in Figure 1B, extracting features that encapsulate important characteristics unique to each cluster.

Our proposed method ClearF++ has several advantages over ClearF and ClearF-one. It is particularly useful when an appropriate number of features must be selected from data with a large number of features, such as in biomarker identification. Additionally, it can be applied when the number of samples is too small compared with the number of features, making it difficult to learn ClearF stably. In summary, ClearF++ divides the data into feature-wise clusters and applies ClearF-one to each cluster, enabling us to select multiple informative features from a large number of features.

3. Results

3.1. Datasets

We conducted an experiment on the gene expression data of lung cancer patients using the ARCHS4 dataset [22], which has been used in several studies [23,24]. We removed genes that had more than 25% zero expression across all samples. The experiment was tested with 8710 genes and 3079 samples. Out of the 3079 samples, 1158 samples belong to the A549 cells (non-small-cell lung cancer) and 1921 samples belong to the IMR90 cells (normal lung fibroblast). Additionally, we performed experiments on several benchmark datasets. To externally validate our results, we further conducted the experiments over two additional benchmark datasets, colon and ALL/AML leukemia datasets [7]. The ALL/AML dataset consists of 72 samples, 47 samples belong to acute lymphoblastic leukemia (ALL) and 25 samples belong to acute myeloid leukemia (AML) [25]. Gene expression levels were measured using Affymetrix high-density oligonucleotide arrays containing 7129 genes. The colon dataset consists of 62 samples, of which 22 are normal and 40 are colon tumor tissue samples [26]. Gene expression levels were measured using Affymetrix oligonucleotide arrays containing expression levels for the 2000 genes with the highest minimal intensity across the samples, as it is prepared in the paper [26].

3.2. *Performance Evaluation on Multiple Benchmark Datasets*

We compared the performance of several feature selection algorithms, including MultiSURF, IFS, and ClearF, to demonstrate the effectiveness of our proposed method in extracting the most relevant features for lung cancer classification using the ARCHS4 dataset. The selected features from each method were used for classification and their AUCs were compared. We performed 10-fold cross-validation by dividing the entire dataset into 10 folds, with one fold for test data and the remaining folds for training data. Each feature selection algorithm was applied solely to the training data to select important features. A classification algorithm was then applied using only the selected features, and the average AUC of the 10-fold cross-validation was measured. The classification model is a basic four-layer DNN, consisting of an input layer, two hidden layers, and an output layer. The sizes of the hidden layers were determined as the number of selected features * 2 and the number of selected features, respectively. The hyperbolic tangent served as the activation function. The Adam optimizer was employed for learning with a learning rate of 1×10^{-3} and 500 epochs using a full batch.

To validate the stability and effectiveness of our proposed method, ClearF++, we conducted a performance comparison using the ARCHS4 lung cancer dataset and several benchmark datasets, such as colon and ALL/AML. The results are displayed in Table 1. These experiments demonstrate that ClearF++ mostly outperforms (p-value < 0.05) other feature selection methods, such as MultiSURF [11], IFS [27], and ClearF [20], across the ARCHS4, colon, and ALL/AML datasets. ClearF++ achieved the highest performance across most of the feature subsets, reaching its best performance at 60 features both in the colon dataset (AUC = 0.826) and the ARCHS4 lung dataset (AUC = 0.983). Likewise, in the ALL/AML dataset, excluding the comparison with ClearF when the number of features was 60, ClearF++ outperformed other methods (p-value < 0.05), achieving the best performance at both 45 and 60 features (AUC = 0.949). Overall, these results highlight the consistent and enhanced performance of ClearF++ across varying numbers of features, showing the robustness and effectiveness of ClearF++. To show the statistical significance of the improvement, we included the results of a paired t-test between ClearF++ and other methods in Table S2 of the Supplementary Material, aligning with the results in Table 1. The results predominantly affirmed the notable superiority of our proposed method compared with other methods, with the exceptions of the case where ClearF++ vs. ClearF selected 60 features in the ALL/AML dataset, and the case where 45 features were selected in the colon dataset.

Table 1. Performance comparison on several benchmark datasets across a varying number of features. The performance was measured with the average AUC of 10-fold cross-validation.

n [1]	Colon				ALL/AML				ARCHS4			
	MultiSURF	IFS	ClearF	ClearF++	MultiSURF	IFS	ClearF	ClearF++	MultiSURF	IFS	ClearF	ClearF++
15	0.648	0.749	0.707	**0.805**	0.906	0.913	0.912	**0.947**	0.925	0.927	0.949	**0.959**
30	0.711	0.672	0.765	**0.773**	0.927	0.926	0.921	**0.940**	0.944	0.953	0.943	**0.973**
45	0.703	0.658	**0.801**	0.751	0.938	0.915	0.927	**0.949**	0.955	0.955	0.952	**0.980**
60	0.723	0.761	0.815	**0.826**	0.927	0.915	**0.949**	**0.949**	0.969	0.949	0.951	**0.983**

[1] The number of features.

3.3. *Performance Evaluation Across Varying Feature and Sample Sizes*

To assess the stability of our proposed method with a limited number of data samples, we evaluated lung cancer classification performance using only 5% of the training data samples from the ARCHS4 dataset. The number of features to be selected increased by 5, starting from 10, in accordance with the experimental procedure employed in the previous study [20].

Figure 2 presents the experimental results on the ARCHS4 lung cancer dataset, using only 5% of the training samples. Our proposed method, ClearF++, exhibited superior performance compared with ClearF and other comparable feature selection algorithms. As displayed in Figure 2A, we observed that the stability of ClearF++ was preserved, while other methods yielded relatively unstable and poor performances with a limited number of samples.

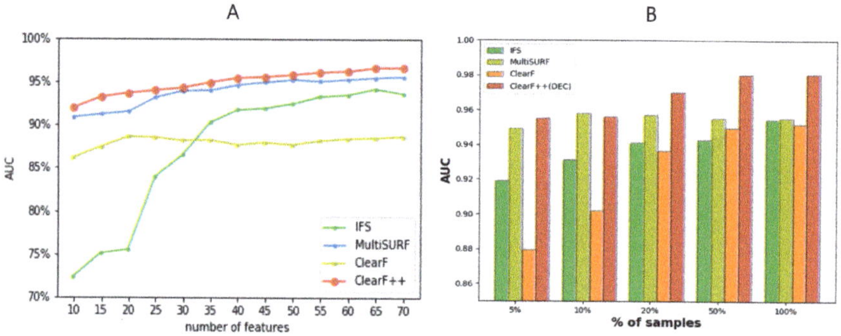

Figure 2. (**A**) Performance comparison across varying numbers of features between ClearF++ and other algorithms. The experiment was obtained using only 5% of the sample for training. (**B**) A comparative experiment measuring performances as the number of samples changes. The experiment was conducted by fixing the number of features at 45.

In Figure 2B, we conducted experiments with varying numbers of samples to investigate the stability of ClearF++ across different sample sizes, fixing the number of features at 45. The experimental results reveal that ClearF++ and MultiSURF showed stable and improved performance across varying numbers of features, even with a small sample size. However, ClearF++ outperformed other methods when more than 10% of the samples were used, whereas MultiSURF exhibited no improvement when larger sample sizes were used and even suffered from slight performance degradation.

These results indicate that our proposed method, ClearF++, demonstrates impressive stability and performance even when dealing with a limited number of data samples. In comparison with other feature selection algorithms, ClearF++ consistently outperforms them, particularly when utilizing more than 10% of the training samples. This highlights the robustness and effectiveness of ClearF++ in various feature or sample size scenarios. These findings emphasize the potential of ClearF++ as a robust and effective feature selection technique, capable of maintaining its performance across a range of feature or sample sizes.

3.4. *Effect of Feature-Wise Clustering Algorithms*

We conducted an ablation study to confirm that our proposed method, ClearF++, indeed contributes to performance improvement compared with the previously proposed method, ClearF. For low-dimensional embedding, KernelPCA with an RBF kernel that showed the best performance in ClearF was utilized in ClearF-one and ClearF++. For the clustering method in ClearF++, k-means and DEC [21] were used. The ablation study was conducted on the ARCHS4 lung cancer dataset, with the results shown in Figure 3. Figure 3A displays the results for the entire samples, while Figure 3B presents the result using only 5% of the samples for training.

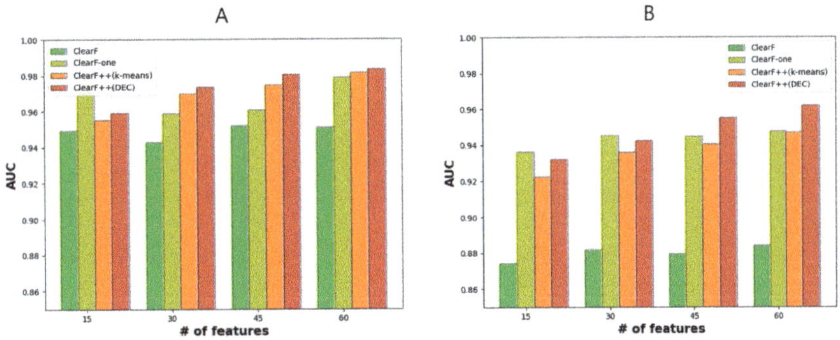

Figure 3. Performance evaluation of ClearF-based methods with various experimental settings in a lung cancer dataset. ClearF-one limits the number of bottleneck layers to 1, and ClearF++ applies feature-wise clustering in the proposed algorithm; thus, two clustering algorithms were compared. (**A**) Results using entire samples. (**B**) Results using 5% of samples for training.

The experimental results show that ClearF-one, which substantially restricts the number of bottleneck layers, contributes to performance improvement, particularly when only a small portion of samples (5%) is used for training. This suggests that ClearF-one is effective in handling limited data samples and can still yield improved performance by simplifying the architecture, reducing the complexity, and focusing on the most relevant features. It is noteworthy that in the context of k-means clustering, the results yielded from training with only 5% of the samples shortly underperformed in comparison with those obtained from ClearF-one without clustering. Under the constraints of a small data size, k-means clustering appeared to struggle in achieving effective clustering. However, when we employed the more sophisticated clustering algorithm, DEC, we observed stable and enhanced performances when feature-wise clustering was applied. Particularly, DEC outperformed k-means clustering in terms of performance when a larger number of features were being selected, both in the scenarios with all samples in Figure 3A and with a small number of samples in Figure 3B. This suggests that DEC may be more suitable for handling complex data structures and capturing underlying patterns in the data when compared with the k-means clustering algorithm, particularly in situations with limited data samples.

We evaluated ClearF++ performance by employing two different clustering methods with varying numbers of clusters. Figure 4A presents the results using the DEC clustering algorithm, while Figure 4B shows the results using k-means clustering. The results reveal that DEC with 15 clusters yielded the best performance, while k-means clustering achieved optimal results with 5 clusters. Although the best performances of these two clustering methods were comparable (close to AUC = 0.98), k-means clustering showed considerable variance depending on the number of clusters. In contrast, DEC demonstrated smaller variance and consistently higher performance across different cluster numbers. Consequently, while both clustering methods (DEC and k-means) can achieve comparable performances, DEC not only improved performances but showed more consistent and reliable results across different numbers of clusters. It suggests that the choice of clustering algorithm is important and that the proposed feature-wise clustering idea of ClearF++ contributes to more robust and improved performances.

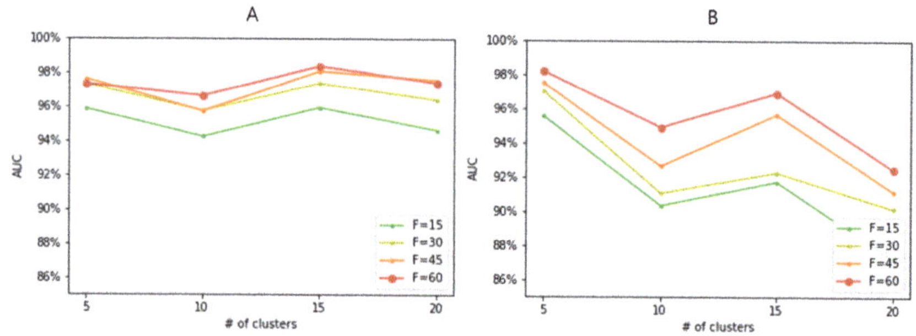

Figure 4. Performance evaluation based on varying numbers of clusters in the lung cancer dataset. F represents the number of selected features. (**A**) Performances of ClearF++ when DEC clustering is applied. (**B**) Performances of ClearF++ when k-means clustering is applied.

3.5. Computational Cost Validation

To verify the computational efficiency of ClearF++ over other methods, we measured and compared the CPU time by running each method 10 times. The experiment utilized data containing 5000 randomly generated features and 5000 samples. As shown in Table 2, ClearF++ showed a substantially faster execution time than MultiSURF and IFS. While it showed a slightly slower execution time than ClearF due to the inclusion of clustering time, the improved performance ensures its competitiveness and effectiveness in comparison with other methods.

Table 2. Computational costs comparison of ClearF++ and other feature selection methods.

Methods	**IFS**	**MultiSURF**	**ClearF**	**ClearF++**
CPU times (s)	398.91 ± 17.22	78,515.59 ± 346.53	132.68 ± 6.51	166.85 ± 10.77

3.6. Functional Enrichment Analysis

To identify high-scoring features, we analyzed the ARCHS4 dataset using ClearF++, which demonstrated improved performance in our experiments. Since scores were calculated for each of the 10 folds, we defined an integrated scoring method. The top 100 features in each fold were assigned scores in descending order, ranging from 100 to 1 point. We then calculated the rank scores by averaging the scores obtained across all folds. The 50 highest-scoring genes are shown in Table S1. Furthermore, to investigate the biological relationships among the selected genes, we performed pathway and gene ontology enrichment analysis using ToppGene [28] on the top 50 genes. The results are shown in Table 3, respectively.

In biomarker detection, high classification accuracy of selected features does not guarantee that features associated with the disease are selected. Considering that the purpose of our algorithm is to select features for identifying important biomarkers, it is crucial to determine whether the top-scoring features are associated with the target disease. The 50 genes with high scores in our method are shown in Table S1. We performed enrichment analysis using these 50 genes.

The enrichment analysis results conducted with ToppGene [28] reveal that several genes related to the glycosaminoglycan (GAG) metabolism pathway received high scores. The extracellular matrix (ECM) regulates cell fate, and glycosaminoglycans (GAGs) are major macromolecules that compose the ECM, which play well-known roles in cancer angiogenesis, proliferation, invasion, and metastasis [29]. GAGs have been widely studied as treatments for cancer, inflammation, infection, and lung diseases, and one study [30] clarified the role of GAGs, contributing to future research.

Table 3. Pathway and gene ontology enrichment analysis results using ToppGene on the top 50 ranked genes. The 10 most significant gene ontology (GO) terms that have the lowest p-values are shown, as well as pathway and disease terms with significant *p*-values ($p < 0.05$) from the enrichment analysis.

Category	ID	Name	*p*-Value	q-Value [1]	q-Value [2]	HC [3]	HCG [4]
BP	GO:0032963	collagen metabolic process	2.89×10^{-8}	5.13×10^{-5}	5.13×10^{-5}	7	144
BP	GO:0030042	actin filament depolymerization	5.03×10^{-7}	8.95×10^{-4}	3.68×10^{-4}	5	71
BP	GO:0032964	collagen biosynthetic process	6.63×10^{-7}	1.18×10^{-3}	3.68×10^{-4}	5	75
MF	GO:0044877	protein-containing complex binding	7.35×10^{-7}	2.09×10^{-4}	2.09×10^{-4}	16	1726
BP	GO:0001568	blood vessel development	8.27×10^{-7}	1.47×10^{-3}	3.68×10^{-4}	13	1152
BP	GO:0035904	aorta development	1.47×10^{-6}	2.62×10^{-3}	4.71×10^{-4}	5	88
BP	GO:0001944	vasculature development	1.87×10^{-6}	3.33×10^{-3}	4.71×10^{-4}	13	1239
BP	GO:0030198	extracellular matrix organization	2.21×10^{-6}	3.93×10^{-3}	4.71×10^{-4}	8	394
BP	GO:0043062	extracellular structure organization	2.25×10^{-6}	4.00×10^{-3}	4.71×10^{-4}	8	395
BP	GO:0045229	external encapsulating structure organization	2.38×10^{-6}	4.24×10^{-3}	4.71×10^{-4}	8	398
Disease	C0268362	Osteogenesis imperfecta type III (disorder)	1.90×10^{-6}	3.42×10^{-3}	3.42×10^{-3}	3	11
Pathway	1269980	Heparan sulfate/heparin (HS-GAG) metabolism	1.82×10^{-5}	5.99×10^{-3}	1.82×10^{-3}	4	54
Pathway	1309217	Defective B3GALT6 causes EDSP2 and SEMDJL1	2.21×10^{-5}	7.26×10^{-3}	1.82×10^{-3}	3	19
Pathway	1269015	Defective B3GAT3 causes JDSSDHD	2.21×10^{-5}	7.26×10^{-3}	1.82×10^{-3}	3	19
Pathway	1269014	Defective B4GALT7 causes EDS, progeroid type	2.21×10^{-5}	7.26×10^{-3}	1.82×10^{-3}	3	19
Pathway	1269981	A tetrasaccharide linker sequence is required for GAG synthesis	5.84×10^{-5}	1.92×10^{-2}	3.20×10^{-3}	3	26
Pathway	1269011	Diseases associated with glycosaminoglycan metabolism	5.84×10^{-5}	1.92×10^{-2}	3.20×10^{-3}	3	26
Pathway	1269982	HS-GAG biosynthesis	9.99×10^{-5}	3.29×10^{-2}	4.69×10^{-3}	3	31
Pathway	M39870	Type I collagen synthesis in the context of osteogenesis imperfecta	1.21×10^{-4}	3.97×10^{-2}	4.97×10^{-3}	3	33
Pathway	1268756	Unfolded Protein Response (UPR)	1.48×10^{-4}	4.88×10^{-2}	5.42×10^{-3}	4	92

[1] Bonferroni q-value, [2] FDR B&H q-value, [3] Hit Count in the query list, [4] Hit count in the genome.

Among the GAG-associated genes that received high scores in our method are GPC1, NDST1, CSPG4, and SDC3. An experiment involving CSPG4-specific mAb 225.28 demonstrated the regression induction of tumor metastasis in a lung metastasis model [31]. Endothelial cell (ECs) junction disassembly, a key step in inflammation, allows for vascular leakage during disease, and thrombin-cleaved fragments of the SDC3 ectodomain promote this process in human lung microvessels in certain cases [32]. NDST1 participates in the synthesis of the heparan sulfate (HS) chain of HSPG, and a study [33] found that it may provide an explanation for the clinical observation that heparin can improve outcomes in small-cell lung cancer (SCLC). Another study [34] suggested that NDST1 is associated with angiogenesis and tumor growth in lung tumors. There is also a study that recommended the use of glypican-1 (GPC1) as an additional positive marker for lung squamous cell carcinoma [35]. These findings suggest that analyzing the effects of NDST1 and SDC3 expression on pulmonary blood vessels in relation to GAGs may be helpful in diagnosing and treating lung cancer.

Additionally, enrichment analysis results using the top 50 genes with high scores (Table 3) reveal that 7 genes related to the collagen metabolic process are included in the biological process. These genes are P3H3, MRC2, MMP14, ENG, EMILIN1, CREB3L1, and COL1A1, with CREB3L1, EMILIN1, and COL1A1 ranking 1st, 3rd, and 4th, respectively. Impaired collagen metabolism is accompanied by increased prolidase activity in lung cancer squamous epithelium [36]. Furthermore, idiopathic pulmonary fibrosis (IPF) is associated with an increased risk of lung cancer with elevated collagen and prolidase activity [36–38]. On the other hand, Prolidase Deficiency (PD) and osteogenesis imperfecta (OI) share similar phenotypes [37]. Notably, our enrichment analysis includes CREB3L1 and COL1A1, which, out of the top 10 genes, are associated with osteogenesis imperfecta type III (disorder). The expression of $\beta 1$ integrin, which has been shown to regulate prolidase activity, is decreased in OI [37,39,40]. However, there is no difference in the levels of $\beta 1$ integrin between healthy lung cells and cancer cells, suggesting that prolidase regulation in lung cancer may involve a different mechanism [36,37]. Therefore, studying the role of prolidase, CREB3L1, and COL1A1 gene expression in lung cancer appears to be significant.

In addition to the aforementioned genes, we found several high-scoring genes (shown in Table S1) that have been linked to lung cancer in multiple studies. The CREB3L1 gene has been associated with lung cancer growth due to its involvement in the activation of alpha-smooth muscle actin (α-SMA)-positive cancer-associated fibroblasts (CAFs) [41]. SYDE1 is associated with epithelial–mesenchymal transition (EMT) reversal, which is associated with the progression of various tumors, including lung cancer [42]. Reduced EMILIN-1 production in some tumor types is associated with higher proliferation of tumor cells in breast and lung cancer [43]. Another study [44] suggested that COL1A1 can be a potential biomarker for poor progression-free survival and chemoresistance in metastatic lung cancer. Serum CKAP4 levels can distinguish lung cancer patients from healthy controls, making it a potential serum diagnostic marker for lung cancer [45]. Carbohydrates associated with LAMP1 play a crucial role in determining lung metastasis [46]. A potential target of TAF15 concerning resistance to radiotherapy, essential for non-small-cell lung cancer treatment, has been proposed [47]. TBX2 subfamily methylation may serve as a potential biomarker for early detection and intervention in non-small-cell lung cancer [48]. Consequently, the genes selected by our method are shown to be biomarker candidates for lung cancer.

4. Discussion

We evaluated the suitability of our methodology for biomarker detection from a machine learning perspective. The results in Figure 2 demonstrate that our proposed method is effective when selecting a small number of features. Particularly, when combined with Figure 3, ClearF-one generally yields favorable results when selecting from 10 to 25 features, and ClearF++ shows improvement when selecting from 30 to 50 features. Given the importance of selecting a small number of features in biomarker discovery, our method can be considered suitable. Moreover, our approach demonstrates stable performance even with a small sample size, as shown in Figure 2B. In the biomedical field, insufficient learning samples are often encountered, and our method proves effective in such cases. Additionally, as shown in Table 2, our method can be effectively employed in environments with limited computational power due to its advantageous execution time.

The results in Figure 3 indicate that ClearF++ can show degraded performances compared with the model without clustering (ClearF-one) when the number of features is small. This is likely because ClearF-one effectively selects a small number of features when only certain information remains after embedding the entirety of the data into a single bottleneck layer. However, when the number of features increases, it suffers from performance degradation due to information loss. In contrast, ClearF++ extracts information for each cluster, which provides more robust and improved performances when selecting multiple features.

Our method addresses the sensitivity issue related to the number of bottleneck layers in the previously proposed method, ClearF, but still requires many parameter adjustments. In particular, determining the number of clusters remains a challenge in clustering. Although Figure 4 shows that our method is not highly sensitive to the number of clusters within a range of 5 to 20 clusters, an exceptionally higher number of clusters, such as 50 or 100, led to instability in clustering results and a substantial performance decrease. Through our experiments, we discerned that the optimal number of clusters likely resides within the 5 to 20 range. However, this range may vary with different datasets according to their sample sizes. Accordingly, future research could focus on the automatic selection of the number of clusters. Further, there is still an issue in setting the model structure or learning method in the part that utilizes DEC. This issue can be addressed in future studies. Furthermore, our method exhibits flexibility towards a range of clustering algorithms. Our experimental findings, as illustrated in Figure 3, indicated that employing a more sophisticated clustering technique could result in more stable and improved performances. Thus, future research that utilizes advanced clustering methods could potentially enhance performances.

5. Conclusions

In this study, we developed an improved feature selection algorithm for identifying biomarkers that can be used for disease prediction and biomedical data analysis. Our experimental results demonstrate several advantages of our method, including improved prediction performance and faster execution Furthermore, it shows substantially stable performance even with a limited number of samples, making it particularly effective for biomedical data analysis, where the available sample size is often insufficient.

One limitation of our method is that it requires determining the optimal number of clusters, which can vary across different datasets. In this study, we experimented with several scenarios to select the most appropriate parameters. However, automatic parameter selection methods can be exploited in future work to address this issue.

Supplementary Materials: The following supporting information can be downloaded at: https://www.mdpi.com/article/10.3390/bioengineering10070824/s1, Table S1: List of the highest-scoring 50 genes in the ARSCH4 lung cancer dataset; Table S2: Results of statistical significance tests between ClearF++ and three other methods: MultiSURF, IFS, and ClearF, corresponding to the results presented in Table 1. The *p*-value was measured through a paired *t*-test between ClearF++ and other methods across three different datasets.

Author Contributions: Conceptualization, S.W.; methodology, S.W. and S.Y.K.; software, S.W.; validation, S.W. and S.Y.K.; formal analysis, S.Y.K. and K.-A.S.; investigation, S.W., S.Y.K. and K.-A.S.; resources, S.W.; writing—original draft preparation, S.W. and S.Y.K.; writing—review and editing, S.Y.K. and K.-A.S.; visualization, S.W. and S.Y.K.; supervision, K.-A.S.; project administration, K.-A.S.; funding acquisition, K.-A.S. All authors have read and agreed to the published version of the manuscript.

Funding: This research was supported by the National Research Foundation of Korea(NRF) grant funded by the Korea government(MSIT)(NRF-2022R1A2C1007434), and also by the Institute of Information & communications Technology Planning & Evaluation (IITP) under the Artificial Intelligence Convergence Innovation Human Resources Development (IITP-2023-RS-2023-00255968) grant funded by the Korea government(MSIT).

Institutional Review Board Statement: Not applicable.

Informed Consent Statement: Not applicable.

Data Availability Statement: All data that are not presented in the main paper are available from the corresponding author on request.

Conflicts of Interest: The authors declare that they have no conflict of interset.

References

1. Group, B.D.W.; Atkinson, A.J., Jr.; Colburn, W.A.; DeGruttola, V.G.; DeMets, D.L.; Downing, G.J.; Hoth, D.F.; Oates, J.A.; Peck, C.C.; Schooley, R.T.; et al. Biomarkers and surrogate endpoints: preferred definitions and conceptual framework. *Clin. Pharmacol. Ther.* **2001**, *69*, 89–95. [CrossRef]
2. Lee, I.H.; Lushington, G.H.; Visvanathan, M. A filter-based feature selection approach for identifying potential biomarkers for lung cancer. *J. Clin. Bioinfor.* **2011**, *1*, 11. [CrossRef]
3. Abeel, T.; Helleputte, T.; Van de Peer, Y.; Dupont, P.; Saeys, Y. Robust biomarker identification for cancer diagnosis with ensemble feature selection methods. *Bioinformatics* **2010**, *26*, 392–398. [CrossRef] [PubMed]
4. Le, T.T.; Blackwood, N.O.; Taroni, J.N.; Fu, W.; Breitenstein, M.K. Integrated machine learning pipeline for aberrant biomarker enrichment (i-mAB): characterizing clusters of differentiation within a compendium of systemic lupus erythematosus patients. *AMIA Annu. Symp. Proc.* **2018**, *2018*, 1358–1367. [PubMed]
5. He, Z.; Yu, W. Stable feature selection for biomarker discovery. *Comput. Biol. Chem.* **2010**, *34*, 215–225. . [CrossRef] [PubMed]
6. Hemphill, E.; Lindsay, J.; Lee, C.; Măndoiu, I.I.; Nelson, C.E. Feature selection and classifier performance on diverse bio- logical datasets. *BMC Bioinform.* **2014**, *15* (Suppl. S13), S4. [CrossRef]
7. Li, J.; Cheng, K.; Wang, S.; Morstatter, F.; Trevino, R.P.; Tang, J.; Liu, H. Feature Selection: A Data Perspective. *ACM Comput. Surv.* **2017**, *50*. [CrossRef]
8. Davis, J.C. *Statistics and Data Analysis in Geology*; John Wiley & Sons, Inc.: Hoboken, NJ, USA, 1973.
9. Hall, M.A. Correlation-Based Feature Selection for Machine Learning. Ph.D. Thesis, The University of Waikato, Hamilton , New Zeland, 1999.

10. Robnik-Šikonja, M.; Kononenko, I. Theoretical and Empirical Analysis of ReliefF and RReliefF. *Mach. Learn.* **2003**, *53*, 23–69. [CrossRef]
11. Urbanowicz, R.J.; Olson, R.S.; Schmitt, P.; Meeker, M.; Moore, J.H. Benchmarking relief-based feature selection methods for bioinformatics data mining. *J. Biomed. Inform.* **2018**, *85*, 168–188. [CrossRef]
12. Peng, H.; Long, F.; Ding, C. Feature selection based on mutual information criteria of max-dependency, max-relevance, and min-redundancy. *IEEE Trans. Pattern Anal. Mach. Intell.* **2005**, *27*, 1226–1238. [CrossRef]
13. Fleuret, F. Fast Binary Feature Selection with Conditional Mutual Information. *J. Mach. Learn. Res.* **2004**, *5*, 1531–1555.
14. Leem, S.; hwan Jeong, H.; Lee, J.; Wee, K.; Sohn, K.A. Fast detection of high-order epistatic interactions in genome-wide association studies using information theoretic measure. *Comput. Biol. Chem.* **2014**, *50*, 19–28. Advances in Bioinformatics: Twelfth Asia Pacific Bioinformatics Conference (APBC2014). [CrossRef]
15. Zhou, H.; Wang, X.; Zhu, R. Feature selection based on mutual information with correlation coefficient. *Appl. Intell.* **2022**, *52*, 5457–5474. [CrossRef]
16. Al-Sarem, M.; Saeed, F.; Alkhammash, E.H.; Alghamdi, N.S. An aggregated mutual information based feature selection with machine learning methods for enhancing iot botnet attack detection. *Sensors* **2022**, *22*, 185. [CrossRef] [PubMed]
17. Cheng, J.; Sun, J.; Yao, K.; Xu, M.; Cao, Y. A variable selection method based on mutual information and variance inflation factor. *Spectrochim. Acta Part A Mol. Biomol. Spectrosc.* **2022**, *268*, 120652. [CrossRef] [PubMed]
18. Hu, L.; Gao, L.; Li, Y.; Zhang, P.; Gao, W. Feature-specific mutual information variation for multi-label feature selection. *Inf. Sci.* **2022**, *593*, 449–471. [CrossRef]
19. Ohyr-Nielsen, M. *Loss of Information by Discretizing Hydrologic Series*; Colorado State University Hydrology Papers; Colorado State University: Fort Collins, CO, USA, 1972.
20. Wang, S.; Jeong, H.H.; Sohn, K.A. ClearF: a supervised feature scoring method to find biomarkers using class-wise embedding and reconstruction. *BMC Med. Genom.* **2019**, *12*, 95. [CrossRef]
21. Xie, J.; Girshick, R.; Farhadi, A. Unsupervised Deep Embedding for Clustering Analysis. In Proceedings of the 33rd International Conference on International Conference on Machine Learning–ICML'16, New York, NY, USA 19–24 June 2016; Volume 48, pp. 478–487.
22. Lachmann, A.; Torre, D.; Keenan, A.B.; Jagodnik, K.M.; Lee, H.J.; Wang, L.; Silverstein, M.C.; Ma'ayan, A. Massive mining of publicly available RNA-seq data from human and mouse. *Nat. Commun.* **2018**, *9*, 1366. [CrossRef]
23. Strobel, B.; Klein, H.; Leparc, G.; Stierstorfer, B.E.; Gantner, F.; Kreuz, S. Time and phenotype-dependent transcriptome analysis in AAV-TGFβ1 and Bleomycin-induced lung fibrosis models. *Sci. Rep.* **2022**, *12*, 12190. [CrossRef]
24. Kaur, N.; Oskotsky, B.; Butte, A.J.; Hu, Z. Systematic identification of ACE2 expression modulators reveals cardiomyopathy as a risk factor for mortality in COVID-19 patients. *Genome Biol.* **2022**, *23*, 15. [CrossRef]
25. Golub, T.R.; Slonim, D.K.; Tamayo, P.; Huard, C.; Gaasenbeek, M.; Mesirov, J.P.; Coller, H.; Loh, M.L.; Downing, J.R.; Caligiuri, M.A.; et al. Molecular classification of cancer: Class discovery and class prediction by gene expression monitoring. *Science* **1999**, *286*, 531–537. [CrossRef] [PubMed]
26. Alon, U.; Barkai, N.; Notterman, D.A.; Gish, K.; Ybarra, S.; Mack, D.; Levine, A.J. Broad patterns of gene expression revealed by clustering analysis of tumor and normal colon tissues probed by oligonucleotide arrays. *Proc. Natl. Acad. Sci. USA* **1999**, *96*, 6745–6750. [CrossRef] [PubMed]
27. Roffo, G.; Melzi, S.; Castellani, U.; Vinciarelli, A.; Cristani, M. Infinite Feature Selection: A Graph-based Feature Filtering Approach. *IEEE Trans. Pattern. Anal. Mach. Intell.* **2021**, *43*, 4396–4410. [CrossRef]
28. Chen, J.; Bardes, E.E.; Aronow, B.J.; Jegga, A.G. ToppGene Suite for gene list enrichment analysis and candidate gene prioritization. *Nucleic Acids Res.* **2009**, *37*, W305–W311. [CrossRef]
29. Wei, J.; Hu, M.; Huang, K.; Lin, S.; Du, H. Roles of Proteoglycans and Glycosaminoglycans in Cancer Development and Progression. *Int. J. Mol. Sci.* **2020**, *21*, 5983. [CrossRef]
30. Morla, S. Glycosaminoglycans and Glycosaminoglycan Mimetics in Cancer and Inflammation. *Int. J. Mol. Sci.* **2019**, *20*, 1963. [CrossRef] [PubMed]
31. Wang, X.; Osada, T.; Wang, Y.; Yu, L.; Sakakura, K.; Katayama, A.; McCarthy, J.B.; Brufsky, A.; Chivukula, M.; Khoury, T.; et al. CSPG4 protein as a new target for the antibody-based immunotherapy of triple-negative breast cancer. *J. Natl. Cancer Inst.* **2010**, *102*, 1496–1512. [CrossRef] [PubMed]
32. Arokiasamy, S.; Balderstone, M.J.M.; De Rossi, G.; Whiteford, J.R. Syndecan-3 in Inflammation and Angiogenesis. *Front. Immunol.* **2019**, *10*, 3031. [CrossRef]
33. Hu, Z.; Wang, C.; Xiao, Y.; Sheng, N.; Chen, Y.; Xu, Y.; Zhang, L.; Mo, W.; Jing, N.; Hu, G. NDST1-dependent heparan sulfate regulates BMP signaling and internalization in lung development. *J. Cell. Sci.* **2009**, *122*, 1145–1154. [CrossRef]
34. Marques, C.; Reis, C.A.; Vivès, R.R.; Magalhães, A. Heparan Sulfate Biosynthesis and Sulfation Profiles as Modulators of Cancer Signalling and Progression. *Front. Oncol.* **2021**, *11*, 778752. [CrossRef]
35. Kai, Y.; Amatya, V.J.; Kushitani, K.; Kambara, T.; Suzuki, R.; Fujii, Y.; Tsutani, Y.; Miyata, Y.; Okada, M.; Takeshima, Y. Glypican-1 is a novel immunohistochemical marker to differentiate poorly differentiated squamous cell carcinoma from solid predominant adenocarcinoma of the lung. *Transl. Lung Cancer Res.* **2021**, *10*, 766–775. [CrossRef] [PubMed]
36. Karna, E.; Surazynski, A.; Palka, J. Collagen metabolism disturbances are accompanied by an increase in prolidase activity in lung carcinoma planoepitheliale. *Int. J. Exp. Pathol.* **2000**, *81*, 341–347. [CrossRef]

37. Eni-Aganga, I.; Lanaghan, Z.M.; Balasubramaniam, M.; Dash, C.; Pandhare, J. PROLIDASE: A Review from Discovery to its Role in Health and Disease. *Front. Mol. Biosci.* **2021**, *8*, 723003. [CrossRef] [PubMed]
38. Ballester, B.; Milara, J.; Cortijo, J. Idiopathic Pulmonary Fibrosis and Lung Cancer: Mechanisms and Molecular Targets. *Int. J. Mol. Sci.* **2019**, *20*, 593. [CrossRef] [PubMed]
39. Galicka, A.; Wolczyñski, S.; Anchim, T.; Surazyñski, A.; Lesniewicz, R.; Palka, J. Defects of type I procollagen metabolism correlated with decrease of prolidase activity in a case of lethal osteogenesis imperfecta. *Eur. J. Biochem.* **2001**, *268*, 2172–2178. [CrossRef]
40. Galicka, A.; Wołczyński, S.; Gindzieński, A.; Surazyński, A.; Pałka, J. Gly511 to Ser substitution in the COL1A1 gene in osteogenesis imperfecta type III patient with increased turnover of collagen. *Mol. Cell Biochem.* **2003**, *248*, 49–56. [CrossRef]
41. Pan, Z.; Xu, T.; Bao, L.; Hu, X.; Jin, T.; Chen, J.; Chen, J.; Qian, Y.; Lu, X.; Li, L.; et al. CREB3L1 promotes tumor growth and metastasis of anaplastic thyroid carcinoma by remodeling the tumor microenvironment. *Mol. Cancer* **2022**, *21*, 190. [CrossRef]
42. Huang, R.Y.J.; Kuay, K.T.; Tan, T.Z.; Asad, M.; Tang, H.M.; Ng, A.H.C.; Ye, J.; Chung, V.Y.; Thiery, J.P. Functional relevance of a six mesenchymal gene signature in epithelial-mesenchymal transition (EMT) reversal by the triple angiokinase inhibitor, nintedanib (BIBF1120). *Oncotarget* **2015**, *6*, 22098–22113. [CrossRef]
43. Amor López, A.; Mazariegos, M.S.; Capuano, A.; Ximénez-Embún, P.; Hergueta-Redondo, M.; Recio, J.Á.; Muñoz, E.; Al-Shahrour, F.; Muñoz, J.; Megías, D.; et al. Inactivation of EMILIN-1 by Proteolysis and Secretion in Small Extracellular Vesicles Favors Melanoma Progression and Metastasis. *Int. J. Mol. Sci.* **2021**, *22*, 7406. [CrossRef]
44. Hou, L.; Lin, T.; Wang, Y.; Liu, B.; Wang, M. Collagen type 1 alpha 1 chain is a novel predictive biomarker of poor progression-free survival and chemoresistance in metastatic lung cancer. *J. Cancer* **2021**, *12*, 5723–5731. [CrossRef]
45. Yanagita, K.; Nagashio, R.; Jiang, S.X.; Kuchitsu, Y.; Hachimura, K.; Ichinoe, M.; Igawa, S.; Fukuda, E.; Goshima, N.; Satoh, Y.; et al. Cytoskeleton-Associated Protein 4 Is a Novel Serodiagnostic Marker for Lung Cancer. *Am. J. Pathol.* **2018**, *188*, 1328–1333. [CrossRef] [PubMed]
46. Agarwal, A.K.; Srinivasan, N.; Godbole, R.; More, S.K.; Budnar, S.; Gude, R.P.; Kalraiya, R.D. Role of tumor cell surface lysosome-associated membrane protein-1 (LAMP1) and its associated carbohydrates in lung metastasis. *J. Cancer Res. Clin. Oncol.* **2015**, *141*, 1563–1574. [CrossRef] [PubMed]
47. Singh, A.K.; Kapoor, V.; Thotala, D.; Hallahan, D.E. TAF15 contributes to the radiation-inducible stress response in cancer. *Oncotarget* **2020**, *11*, 2647–2659. [CrossRef] [PubMed]
48. Nehme, E.; Rahal, Z.; Sinjab, A.; Khalil, A.; Chami, H.; Nemer, G.; Kadara, H. Epigenetic Suppression of the T-box Subfamily 2 (TBX2) in Human Non-Small Cell Lung Cancer. *Int. J. Mol. Sci.* **2019**, *20*, 1159. [CrossRef] [PubMed]

bioengineering

MDPI

Article

Dense Multi-Scale Graph Convolutional Network for Knee Joint Cartilage Segmentation

Christos Chadoulos [1], Dimitrios Tsaopoulos [2], Andreas Symeonidis [1], Serafeim Moustakidis [1] and John Theocharis [1,*]

[1] Department of Electrical & Computer Engineering, Aristotle University of Thessaloniki, 54124 Thessaloniki, Greece; christgc@auth.gr (C.C.); symeonid@ece.auth.gr (A.S.); s.moustakidis@aideas.eu (S.M.)
[2] Institute for Bio-Economy and Agri-Technology, Centre for Research and Technology—Hellas, 38333 Volos, Greece; d.tsaopoulos@certh.gr
* Correspondence: theochar@ece.auth.gr

Abstract: In this paper, we propose a dense multi-scale adaptive graph convolutional network (*DMA-GCN*) method for automatic segmentation of the knee joint cartilage from MR images. Under the multi-atlas setting, the suggested approach exhibits several novelties, as described in the following. First, our models integrate both local-level and global-level learning simultaneously. The local learning task aggregates spatial contextual information from aligned spatial neighborhoods of nodes, at multiple scales, while global learning explores pairwise affinities between nodes, located globally at different positions in the image. We propose two different structures of building models, whereby the local and global convolutional units are combined by following an alternating or a sequential manner. Secondly, based on the previous models, we develop the *DMA-GCN* network, by utilizing a densely connected architecture with residual skip connections. This is a deeper *GCN* structure, expanded over different block layers, thus being capable of providing more expressive node feature representations. Third, all units pertaining to the overall network are equipped with their individual adaptive graph learning mechanism, which allows the graph structures to be automatically learned during training. The proposed cartilage segmentation method is evaluated on the entire publicly available Osteoarthritis Initiative (*OAI*) cohort. To this end, we have devised a thorough experimental setup, with the goal of investigating the effect of several factors of our approach on the classification rates. Furthermore, we present exhaustive comparative results, considering traditional existing methods, six deep learning segmentation methods, and seven graph-based convolution methods, including the currently most representative models from this field. The obtained results demonstrate that the *DMA-GCN* outperforms all competing methods across all evaluation measures, providing $DSC = 95.71\%$ and $DSC = 94.02\%$ for the segmentation of femoral and tibial cartilage, respectively.

Keywords: knee cartilage osteoarthritis (*KOA*); magnetic resonance imaging (*MRI*) segmentation; multi-atlas; graph neural networks (*GNNs*); deep learning; graph learning; semi-supervised learning (*SSL*)

Citation: Chadoulos, C.; Tsaopoulos, D.; Symeonidis, A.; Moustakidis, S.; Theocharis, J. Dense Multi-Scale Graph Convolutional Network for Knee Joint Cartilage Segmentation. *Bioengineering* **2024**, *11*, 278. https://doi.org/10.3390/bioengineering11030278

Academic Editors: Hongqing Yu, Alaa AlZoubi, Yifan Zhao and Hongbo Du

Received: 14 February 2024
Revised: 7 March 2024
Accepted: 11 March 2024
Published: 14 March 2024

1. Introduction

Osteoarthritis (OA) is one of the most prevalent joint diseases worldwide, causing pain and mobility issues, reducing the ability to lead an independent lifestyle, and ultimately decreasing the quality of life in patients. It primarily manifests among populations of advanced age, with an estimated 10% of people over the age of 55 dealing with this condition. That percentage is likely to noticeably increase in the coming years, especially in the developing parts of the world where life expectancy is steadily on the rise [1].

Among the available imaging modalities, magnetic resonance imaging (MRI) constitutes a valuable tool in the characterization of the knee joint, providing a robust quantitative and qualitative analysis for detecting anatomical changes and defects in the cartilage tissue. Unfortunately, manual delineations performed by human experts are resource- and time-consuming, while also suffering from unacceptable levels of inter- and intrarater variability. Thus, there is an increasing demand for accurate and time-efficient fully automated methods for achieving reliable segmentation results.

During the past decades, a considerable amount of research has been conducted to achieve the above goals. However, the thin cartilage structure, as well as the great variability in and intensity inhomogeneity of MR images have posed significant challenges. Several methods are proposed to address those issues, ranging from more traditional image processing ones such as statistical shape models and active appearance models, to more automated ones employing classical machine learning and deep learning techniques. A comprehensive review of such automated methods for knee articular segmentation can be found in [2].

1.1. Statistical Shape Methods

A wide variety of methods that fall under the statistical shape model (*SSM*) and active appearance model (*AAM*) family have been extensively employed in knee joint segmentation applications in the past. Since the specific shape of the cartilage structure is quite distinct and characteristic, these methods employ this feature as a stepping stone towards complete delineation for the whole knee joint [3]. Additionally, *SSMs* have been successfully utilized as shape regularizers within more complex segmentation pipelines, mainly as a final postprocessing step [4]. While conceptually simple, these methods are highly sensitive to the initial landmark selection process.

1.2. Machine Learning Methods

Under the classical machine learning setting, knee cartilage segmentation is cast as a supervised classification task, estimating the label of each voxel from a set of handcrafted or automatically extracted features from the available set of images. Typical examples of such approaches can be found in [5,6]. These methods are conceptually simple but usually offer mediocre results, due to their poor generalization capabilities and the utilization of fixed feature descriptors that may not be well suited to efficiently capture the data variability.

1.3. Multi-Atlas Patch-Based Segmentation Methods

Multi-atlas patch-based methods have long been a staple in medical imaging applications [7,8]. Utilizing an atlas library $\mathcal{A} = \{\mathcal{A}_i, \mathcal{L}_i\}_{i=1}^{n_\mathcal{A}}$ comprising $n_\mathcal{A}$ magnetic resonance images ($\mathcal{A}_i$) and their corresponding label maps $\mathcal{L}_i$, these methods operate on a single target image $\mathcal{T}$ at a time, annotating it by propagating voxel labels from the atlas library $\mathcal{A}$. The implicit assumption in this framework is that the target image $\mathcal{T}$ and the corresponding images comprising the atlas library $\mathcal{A}$ reside in a common coordinate space. This assumption is enforced by registering all atlases $\mathcal{A}_i \in \mathcal{A}$ along with their corresponding labels maps $\mathcal{L}_i$ to the target image space, via an affine or deformable transformation [9].

Multi-atlas patch-based methods usually consist of the following steps: For all voxels $\mathbf{x} \in \mathcal{T}$, a search volume $N(\mathbf{x})$ of size $N_s = (n \times n \times n)$ centered around $\mathbf{x}$ is formed, and every corresponding voxel $\mathbf{y} \in N(\mathbf{x})$ in the spatially adjacent locations in the registered atlases yields a patch library $\mathbf{P}_\mathcal{L} = \mathbf{p}_{\mathcal{A}_i}(\mathbf{y}), \forall \mathbf{y} \in N(\mathbf{x})$ for all atlases $i = 1, \cdots, n_\mathcal{A}$. An optimization problem such as sparse coding (*SC*) is then used to reconstruct the target patch as a linear combination of its corresponding atlas library. Established methods of this category of segmentation algorithms are presented in [7,8]. Despite being capable of achieving appreciable results, these methods do not scale well to large datasets due to the intense computational demands of constructing a patch library and solving an optimization problem for each voxel of every target image.

1.4. Deep Learning Methods

The recent resurgence of deep learning has had a great impact on medical imaging applications, with an increasing number of works reporting the use of deep architectures in various applications pertaining to that field. Initially restricted to 2D models due to the large computational load imposed by the 3D structure of magnetic resonance images, the recent advancements in processing power have allowed for a wide variety of fully 3D models to be proposed, offering markedly better performance with respect to the more traditional methods [10,11]. In our study, we compare our proposed method against a series of representative deep architectures, with applications ranging from semantic segmentation to point-cloud classification and medical image segmentation. In particular, we consider the *SegNet* [12], *DenseVoxNet* [13], *VoxResNet* [14], *PointNet* [15], *CAN3D* [13], and *KCB-Net* [16] architectures (Section 9.4.2).

1.5. Graph Convolutional Neural Networks

Recently, intensive research has been conducted in the field of graph convolutional networks, owing to their efficiency in handling non-Euclidean data [17]. These models can be distinguished into two general categories, namely, the spectral-based [18,19] and the spatial-based methods [20–24]. The spectral-based networks rely on the graph signal processing principles, utilizing filters to define the node convolutions. The *ChebNet* in [18] approximates the convolutional filters by Chebyshev polynomials of the diagonal matrix of eigenvalues while the *GCN* model in [19] performs a first-order approximation of *ChebNet*.

Spatial-based graph convolutions, on the other hand, update a central node's representation by aggregating the representations of its neighboring nodes. The message-passing neural network (*MPNN*) [20] considers graph convolutions as a message-passing process, whereby information is traversed between nodes via the graph edges. *GraphSAGE* [25] applies node sampling to obtain a fixed number of neighbors for each node's aggregation. A graph attention network (*GAT*) [24] assumes that the contribution of the neighboring nodes to the central one is determined according to a relative weight of importance, a task achieved via a shared attention mechanism across nodes with learnable parameters.

During the last years, *GCN*s have found extensive use in a diverse range of applications, including citation and social networks [19,26], graph-to-sequence learning tasks in natural language processing [27], molecular/compound graphs [28], and action recognition [29]. Considerable research has been conducted on the classification of remotely sensed hyperspectral images [30–32], mainly due to the capabilities of *GCN*s to capture both the spatial contextual information of pixels, as well as the long-range relationships of distant pixels in the image. Another domain of application is the forecasting of traffic features in smart transportation networks [33,34]. To capture the varying spatio-temporal relationships between nodes, integrated models are developed in these works, which combine graph-based spatial convolutions with temporal convolutions.

Finally, to confront the gradient vanishing effect faced by traditional graph-based models, deep *GCN* networks have recently been suggested [35,36]. Particularly, in [35], a densely connected graph convolutional network (*DCGCN*) is proposed for graph-to-sequence learning, which can capture both local and nonlocal features. In addition, Ref. [36] presents a densely connected block of *GCN* layers, which is used to generate effective shape descriptors from 3D meshes of images.

1.6. Outline of Proposed Method

The existing patch-based methods exhibit several drawbacks which can potentially degrade their segmentation performance. First, for each target voxel, these methods construct a local patch library at a specific spatial scale, comprising neighboring voxels from atlas images. Then, classifiers are developed by considering pairwise similarities between voxels in that local region. This suggests that target labeling is accomplished by relying solely on local learning while disregarding the global contextual information among pixels. Hence, long-range relationships among distant voxels in the region of interest are ignored, although these voxels may belong to the same class but with a different textural appearance. Secondly, previous methods in the field resort to inductive learning to produce voxel segmentation, which implies that the features of the unlabeled target voxels are not leveraged during the labeling process. Finally, some recent segmentation methods employ graph-based approaches allowing a more effective description of voxel pairwise affinities via sparse code reconstructions [8,37]. Despite the better data representation, the target voxel labeling is achieved using linear aggregation rules for transferring the atlas voxel's labels, such as the traditional label propagation (*LP*) mechanism via the graph edges. Such first order methods may fail to adequately capture the full scope of dependencies among the voxel representations. The labeling of each voxel proceeds by aggregating spectral information strictly from its immediate neighborhood, failing to exploit long-term dependencies with potentially more similar patches in distant regions of the image, thus ultimately yielding suboptimal segmentation results.

To properly address the above shortcomings, in this paper we present a novel method for the automatic segmentation of knee articular cartilage, based on recent advances in the field of graph-based neural networks. More concretely, we propose the dense multi-scale adaptive graph convolutional network (*DMA-GCN*) method, which constructively integrates local spatial-level learning and global-level contextual learning concurrently. Our goal is to generate, via automatic convolutional learning, expressive node representations by merging pairwise importance at multiple spatial scales with long-range dependencies among nodes for enhanced volume segmentation. We approach the segmentation task as a multi-class classification problem with the five classes: background: 0; femoral bone: 1; femoral cartilage: 2; tibial bone: 3; tibial cartilage: 4. Recognizing the more crucial role of the cartilage structure in the assessment of the knee joint and considering the increased difficulty for its automatic segmentation as contrasted with that of bones, our efforts are primarily devoted to that issue. Figure 1 depicts a schematic framework of the proposed approach. The main properties and innovations of the *DMA-GCN* model are described as follows:

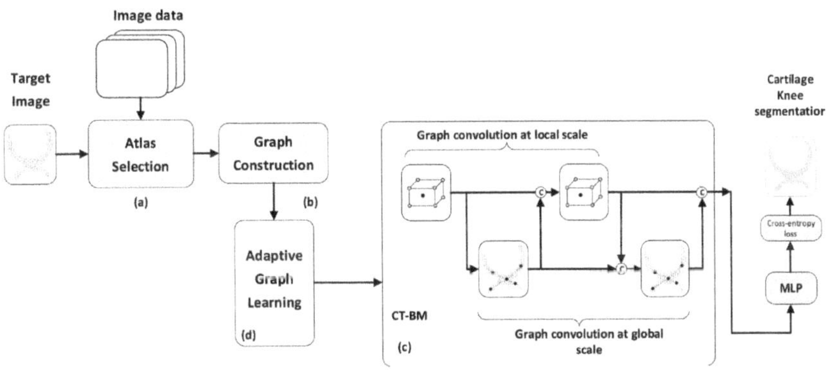

Figure 1. Outline of the proposed knee cartilage segmentation approach. It comprises the atlas subset selection (**a**), the graph construction part (**b**), a specific form of graph-based convolutional model (**c**), the adaptive graph learning (**d**), and the *MLP* network providing the class estimates for the segmentation of the target image. Black dots correspond to central nodes and colored nodes to neighboring ones, respectively. The encircled C symbol represents an aggregation function.

Multi-atlas setting: Our scheme is tailored to the multi-atlas approach, whereby label information from atlas images (labeled) is transferred to segment the target image (unlabeled). To this end, at the preliminary stage, images are aligned using a cost-effective affine registration. Subsequently, for each target image $\mathcal{T}$, we generate its corresponding atlas library $\{\mathcal{A}, \mathcal{L}\}$ according to a similarity criterion (Figure 1a).

Graph construction: This part refers to the way in which images are represented in terms of nodes and the organization of node data to construct the overall graph (Figure 1b). Here, the graph node corresponds to a generic patch of size $5 \times 5 \times 5$ around a central voxel, while the node feature vector is provided by a 3D-HOG feature descriptor. Accordingly, the image is represented as a collection of spatially stratified nodes, covering adequately all classes across the region of interest. Following the multi-atlas setting, we construct sequences of aligned data, comprising target nodes and those for the atlases at spatially correspondent locations. Given the node sequences, we further generate the sequence libraries which are composed of neighboring nodes at various spatial scales. The collection of all node libraries forms the overall graph structure, whereby both local (spatially neighboring) and global (spatially distant) node relationships are incorporated.

Semi-supervised learning (SSL): Following the *SSL* scenario, the input graph data comprise both labeled nodes from the atlas library and unlabeled ones from the target image to be segmented. In that respect, contrary to some existing methods, the features of unlabeled data are leveraged via learning to compute the node embbeddings and label the target nodes.

Local–global learning: As can be seen from Figure 1c, graph convolutions over the layers proceed along two directions, namely, the local spatial level and the global level, respectively. The local spatial branch includes the so-called local convolutional (*Lconv*) units which operate on the subgraph of aligned neighborhoods of nodes (sequence libraries). The node embeddings generated by these units incorporate the contextual information between nodes at a local spatial level. To further improve local search, we integrate local convolutions at multiple scales, so that the local context around nodes is captured more efficiently. On the other hand, the global branch includes global convolution (*Gconv*) units. These units provide the global node embeddings by taking into consideration the pairwise affinities of distant nodes distributed over the entire region of interest of the cartilage volume. The final node representations are then obtained by aggregating the embeddings computed at the local spatial and the global levels, respectively.

Convolutional building models: An important issue is how the local and global hidden representations of nodes are combined across the convolution layers. In this context, we propose two different structures: the cross-talk building model (*CT-BM*) and the sequential building model (*SEQ-BM*). Both models comprise four convolutional units overall, specifically, two *Lconv* and two *Gconv* units, undertaking local and global convolutions, respectively. The *CT-BM* (Figure 1c) performs intertwined local–global learning, with skip connections and aggregators. The links indicate the cross-talks between the two paths. The *SEQ-BM*, on the other hand, adopts a sequential learning scheme. In particular, local spatial learning is completed first, followed by the respective convolutions at the global level.

Adaptive graph learning (AGL): Considering fixed graphs with predetermined adjacency weights among nodes can degrade the segmentation results. To confront this drawback, every *Lconv* and *Gconv* unit is equipped here with an *AGL* mechanism, which allows us to automatically learn the proper graph structure at each layer. At the local spatial level, *AGL* adaptively designates the connectivity relationships between nodes via learnable attention coefficients. Hence, *Lconv* can concentrate and aggregate features from relevant nodes in the local search region. Further, at the global level we propose a different *AGL* scheme for *Gconv* units, whereby graph edges are learned from the input features of each layer.

Densely connected GCN: The proposed *CT-BM* and *SEQ-BM* can be utilized as standalone models to undertake the graph convolution task. Nevertheless, their depth is confined to two local–global layers, since an attempt to deepen the networks is hindered by the gradient-vanishing effect. To circumvent this deficiency, we finally propose a densely connected convolutional network, the *DMA-GCN* model. The *DMA-GCN* considers *CT-BM* or *SEQ-BM* as the building block of the deep structure. It exhibits a deep architecture with skip connections whereby each layer in the block receives feature maps from all previous layers and transmits its outputs to all subsequent layers. Overall, the *DMA-GCN* shares some salient qualities, such as a deep structure with an enhanced performance rate and better information flow, local–global level convolutions, and adaptive graph learning.

In summary, the main contributions of this paper are described as follows.

- A novel multi-atlas approach is presented for knee cartilage segmentation from *MRI* images based on graph convolutional networks which operates under the semi-supervised learning paradigm.
- With the aim to generate expressive node representations, we propose a new learning scheme that integrates graph information at both local and global levels concurrently. The local branch exploits the relevant spatial information of neighboring nodes at multiple scales, while the global branch incorporates global contextual relationships among distant nodes.
- We propose two convolutional building models, the *CT-BM* and *SEQ-BM*. In the *CT-BM*, the local and global learning tasks are intertwined along the layer convolutions, while the *SEQ-BM* follows a sequential mode.
- Both local and global convolutional units, at each layer, are equipped with suitable attention mechanisms, which allows the network to automatically learn the graph connective relationships among nodes during training.
- Using the proposed *CT-BM* and *SEQ-BM* as block units, we finally present a novel densely connected model, the *DMA-GCN*. The network exhibits a deeper structure which leads to more enhanced segmentation results, while at the same time, it shares all salient properties of our approach.
- We have devised a thorough experimental setup to investigate the capabilities of the suggested segmentation framework. In this setting, we examine different test cases and provide an extensive comparative analysis with other segmentation methods.

The remainder of this paper is organized as follows. Section 2 reviews some representative forms of graph convolutional networks related to our work and involved in the experimental analysis. Section 3 presents the image preprocessing steps and the atlas selection process. In Section 4, we discuss the node feature descriptor, as well as the graph construction of the images. Section 5 elaborates on the proposed local and global convolutional units, along with their attention mechanisms. Section 6 describes the suggested convolutional building blocks, while Section 7 presents our densely connected network. Section 8 discusses the transductive vs. inductive learning and the full-batch vs. mini-batch learning in our approach. In Sections 9 and 10, we provide the experimental setup and respective comparative results of the proposed methodology, while Section 11 concludes this study.

2. Related Work

In this section, we review some representative models in the field of graph convolutional networks that are related to our work and are also included in the experiments.

Definition 1. *A graph $\mathcal{G}$ is defined as $\mathcal{G} = (\mathcal{V}, \mathcal{E}, \mathbf{X})$ where $\mathcal{V}$ denotes the set of N nodes $v_i \in \mathcal{V}$, and $\mathcal{E}$ is the set of edges connecting the nodes $(v_i, v_j) \in \mathcal{E}$. $\mathbf{X} \in \mathbb{R}^{N \times F}$ is a matrix subsuming the node feature descriptors $\mathbf{x}_i \in \mathbb{R}^D, i = 1, \ldots N$ with F denoting the feature vector dimensionality. The graph is associated with an adjacency matrix $\boldsymbol{A} \in \mathbb{R}^{N \times N}$ (binary or weighted), which includes the connection links between nodes. A larger entry $\boldsymbol{A}_{ij} > 0$ suggests the existence of a strong relationship between nodes (v_i, v_j), while $\boldsymbol{A}_{ij} = 0$ signifies the lack of connectivity. The graph Laplacian matrix $\boldsymbol{L} \in \mathbb{R}^{(N \times N)}$ is defined as $\boldsymbol{L} = \boldsymbol{D} - \boldsymbol{A}$, where $\boldsymbol{D}$ is the diagonal degree matrix with $\boldsymbol{D}_{ii} = \sum_{j=1}^{N} \boldsymbol{A}_{ij}, i = 1, \ldots, N$. Finally, the normalized graph adjacency matrix with the added self-connections is denoted by $\widetilde{\boldsymbol{A}} = \boldsymbol{A} + \boldsymbol{I}_N$, with the corresponding degree matrix given by $\widetilde{\mathbf{D}}_{ii} = \sum_{j=1}^{N} \widetilde{\boldsymbol{A}}_{ij}$*

2.1. Graph Convolutional Network (GCN)

The *GCN* proposed in [19] is a spectral convolutional model. It tackles the node classification task under the semi-supervised framework, i.e., where labels are available only for a portion of the nodes in the graph. Under this setting, learning is achieved by enforcing a graph Laplacian regularization term with the aim of smoothing the node labels:

$$\mathcal{L} = \mathcal{L}_0 + \lambda \mathcal{L}_{reg} \tag{1}$$

$$\mathcal{L}_{reg} = \sum_i \sum_j \mathbf{A}_{ij} \| f(\mathbf{x}_i) - f(\mathbf{x}_j) \| = f(\mathbf{X})^T \boldsymbol{L} f(\mathbf{X}) \tag{2}$$

where $\mathcal{L}_0$ represents the supervised loss measured on the labeled nodes of the graph, $f(\mathbf{X}, \mathbf{A})$ is a differentiable function implemented by a graph neural network, λ is a regularization term balancing the supervised loss in regard to the overall smoothness of the graph, $\mathbf{X}$ is the node feature matrix, and $\boldsymbol{L}$ is the graph Laplacian.

In a standard multilayer graph-based neural network framework, information flows across the nodes by applying the following layerwise propagation rule:

$$\mathbf{H}^{(l+1)} = \sigma\left(\widetilde{\mathbf{D}}^{-\frac{1}{2}} \widetilde{\mathbf{A}} \widetilde{\mathbf{D}}^{-\frac{1}{2}} \mathbf{H}^{(l)} \mathbf{W}^{(l)}\right) \tag{3}$$

where $\sigma(\cdot)$ denotes the *LeakyReLU*$(\cdot)$ activation function, $W^{(l)}$ is a layer-specific trainable weight matrix, and $\mathbf{H}^{(l)}$ is the matrix of activation functions in the lth layer, with $\mathbf{H}^{(0)} = \mathbf{X}$. The authors in [19] show that the propagation rule in Equation (3) provides a first-order approximation of localized spectral filters on graphs. Most importantly, we can construct multilayered graph convolution networks by stacking several convolutional layers of the form in Equation (3). For instance, a two-layered *GCN* can be represented by

$$\mathbf{Z} = f(\mathbf{X}, \mathbf{A}) = softmax\left(\widehat{\mathbf{A}} ReLU(\widehat{\mathbf{A}} \mathbf{X} \mathbf{W}^{(0)}) \mathbf{W}^{(1)}\right) \tag{4}$$

where $\mathbf{Z}$ is the network's output, $softmax(\cdot)$ is the output layer activation function for multi-class problems, and $\widehat{\mathbf{A}} = \widetilde{\mathbf{D}}^{-\frac{1}{2}} \widetilde{\mathbf{A}} \widetilde{\mathbf{D}}^{-\frac{1}{2}}$ is the normalized adjacency matrix. The weight matrices $\mathbf{W}^{(0)}$ and $\mathbf{W}^{(1)}$ are trained using some variant of gradient descent with the aid of a loss function.

2.2. Graph Attention Network (GAT)

A salient component in *GATs* [24] is an attention mechanism incorporated in the aggregation of the graph attention layers (*GALs*), with the aim to automatically capture valuable relationships between neighboring nodes. Let $\mathbf{H} = \{\mathbf{h}_1, \ldots, \mathbf{h}_N\}, \mathbf{h}_i \in \mathcal{R}^D$ and $\widetilde{\mathbf{H}} = \{\widetilde{\mathbf{h}}_1, \ldots, \widetilde{\mathbf{h}}_N\}, \widetilde{\mathbf{h}}_i \in \mathbb{R}^{\widetilde{D}}$ denote the inputs and outputs of a *GAL*, where N is the number of nodes, while D and $\widetilde{D}$ are the corresponding dimensionalities of the node feature vectors. The convolution process entails three distinct issues: the shared node embeddings, the attention mechanism, and the update of node representations. As an initial step, a learnable transformation parameterized by the weight matrix $\mathbf{W} \in \mathbb{R}^{\widetilde{D} \times D}$ is applied on nodes, with the goal of producing expressive feature representations. Next, for every node pair, a shared attention mechanism is performed on the transformed features,

$$g_{ij} = \boldsymbol{\alpha}^T \cdot (\mathbf{W}\mathbf{h}_i \,||\, \mathbf{W}\mathbf{h}_j) \tag{5}$$

where g_{ij} signifies the importance between nodes $\mathbf{h}_i$ and $\mathbf{h}_j$, $\boldsymbol{\alpha}$ is a learnable weight vector, and $||$ denotes the concatenation operator. To make the above mechanism effective, the computation of the attention coefficients is confined between each node $-i$ and its neighboring nodes $-j$, $j \in \mathcal{N}_i$. In the *GAT* framework, the attention mechanism is implemented by a single-layer feed-forward neural network, parameterized by *LeakyReLU* nonlinearities, which provide the normalized attention coefficients:

$$a_{ij} = \frac{\exp\left(LeakyReLU\left(\boldsymbol{\alpha}^T \cdot [\mathbf{W}\mathbf{h}_i || \mathbf{W}\mathbf{h}_j]\right)\right)}{\sum_{k \in \mathcal{N}_i} \exp\left(LeakyReLU(\boldsymbol{\alpha}^T \cdot [\mathbf{W}\mathbf{h}_i || \mathbf{W}\mathbf{h}_k])\right)} \tag{6}$$

Given the attention coefficients, the node feature representations at the output of the *GAT* are updated via a linear aggregation of neighboring nodes' features

$$\widetilde{\mathbf{h}}_i = \sigma\left(\sum_{j\in\mathcal{N}_i} \alpha_{ij}\mathbf{W}\mathbf{h}_j\right) \tag{7}$$

To stabilize the learning procedure, the previous approach is extended in a *GAT* by considering multiple attention heads. In that case, the node features are computed by

$$\widetilde{\mathbf{h}}_i = \Big\Vert_{k=1}^{K} \sigma\left(\sum_{j\in\mathcal{N}_i} \alpha_{ij}^{(k)}\mathbf{W}^{(k)}\mathbf{h}_j\right) \tag{8}$$

where K is the number of independent attention heads applied, while $\alpha_{ij}^{(k)}$ and $\mathbf{W}^{(k)}$ denote the normalized attention coefficients associated with the kth attention head and its corresponding embedding matrix, respectively. In this work, we exploit the principles of the *GAT*-based attention mechanism in the proposed local convolutional units, with the goal to aggregate valuable contextual information from local neighborhoods, at multiple search scales (Section 5.1).

2.3. *GraphSAGE*

The *GraphSAGE* network in [25] tackles the inductive learning problem, where labels must be generated for previously unseen nodes, or even entirely new subgraphs. *GraphSAGE* aims to learn a set of aggregator functions $AGGR_k, k = 1, \ldots, K$, which are used to aggregate information from each node's local neighborhood. Node aggregation is carried out at multiple spatial scales (hops). Among the different schemes proposed in [25], in our experiments, we consider the max-pooling aggregator, where each neighbor's vector is independently supplied to a fully connected neural network:

$$\mathbf{h}_{\mathcal{N}_v}^{k} = AGGR_k^{pool} = \max\left(\left\{\sigma\left(\mathbf{W}_{pool}\mathbf{h}_{u_i}^{k} + \mathbf{b}\right), \forall u_i \in \mathcal{N}(v)\right\}\right) \tag{9}$$

where $max(\cdot)$ denotes the element-wise max operator, $\mathbf{W}_{pool}$ is the weight matrix of learnable parameters, $\boldsymbol{b}$ is the bias vector, and $\sigma(\cdot)$ is a nonlinear activation function. Further, $\mathbf{h}_{\mathcal{N}(v)}^{k}$ denotes the result obtained after a max-pooling aggregation on the neighboring nodes of node v. *GraphSAGE* then concatenates the current node's representation $\mathbf{h}_v^{k-1}$ with the aggregated neighborhood feature vector $\mathbf{h}_{\mathcal{N}(v)}^{k-1}$ to compute, via a fully connected layer, the updated node feature representations:

$$\mathbf{h}_v^{k} = \sigma\left(\mathbf{W}^{k} \cdot \left(\mathbf{h}_v^{k-1} \,\Vert\, \mathbf{h}_{\mathcal{N}_v}^{k-1}\right)\right) \tag{10}$$

where $\mathbf{W}^{k}$ is a weight matrix associated with aggregator $AGGR_k^{pool}$.

2.4. *GraphSAINT*

GraphSAINT [21] differs from the previously examined architectures in that instead of building a full *GCN* on all the available training data, it samples the training graph itself, creating subsets of the original graph, building and training the associated *GCN*s on those subgraphs. For each mini-batch sampled in this iterative process, a subgraph $\mathcal{G}_s = (V_s, \mathcal{E}_s)$ (where $|V_s| \ll |V|$) is used to construct a *GCN*. Forward and backward propagation is performed, updating the node representations and the participating edge weights. An initial preprocessing step is required for the smooth operation of the process, whereby an appropriate probability of sampling must be assigned to each node and edge of the initial graph.

3. Materials

In this section, we present the dataset used in this study, the image preprocessing steps, and finally, the construction of the atlas library.

3.1. Image Dataset

The MR images used in this study comprise the entirety of the publicly available, baseline Osteoarthritis Initiative (*OAI*) repository, for which segmentation masks are available, consisting of a total of 507 subjects. The specific MRI modality utilized across all the experiments corresponds to the sagittal 3D dual-echo steady-state (*3D-DESS*) sequence with water excitation, with an image size of $384 \times 384 \times 384$ voxels and a voxel size of $0.36 \times 0.36 \times 0.70$ mm. The respective segmentation masks serving as the ground truth are provided by the publicly available repository assembled by [38], including labels for the following knee joint structures (classes): background tissue, femoral bone (*FB*), femoral cartilage (*FC*), tibial bone (*TB*), and tibial cartilage (*TC*). Figure 2 showcases a typical knee MRI, in the three standard orthogonal planes (sagittal, coronal, axial).

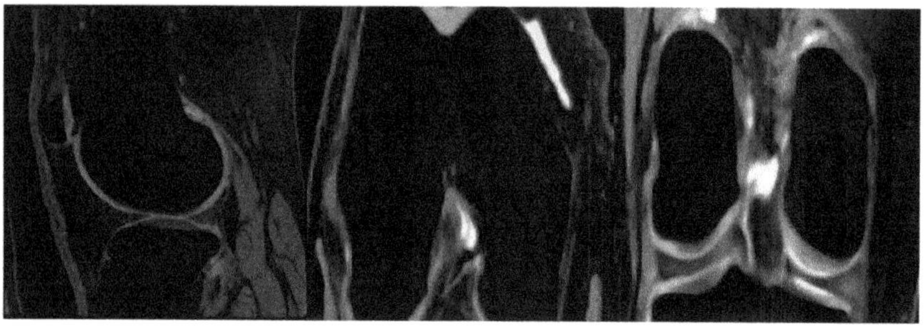

Figure 2. A typical knee MRI viewed in three orthogonal planes (**left** to **right**: sagittal, coronal, axial).

3.2. Image Preprocessing

The primary source of difficulties in automated cartilage segmentation stems from the similar texture and intensity profile of articular cartilage and background tissues, as they are depicted in most MRI modalities, a problem further accentuated by the usually high intersubject variability present in the imaging data. To this end, the images were preprocessed by applying the following steps:

1. *Curvature flow filtering:* A denoising curvature flow filter [39] was applied, with the aim of smoothing the homogeneous image regions, while simultaneously leaving the surface boundaries intact.
2. *Inhomogeneity correction: N3* intensity nonuniform bias field correction [40] was performed on all images, dealing with the issue of intrasubject variability within similar classes among subjects.
3. *Intensity standardization:* MRI histograms were mapped to a common template, as described in [41], ensuring that all associated structures across the subjects shared a similar intensity profile.
4. *Nonlocal-means filtering:* A final filtering process smoothed out any leftover artefacts and further reduced noise. The method presented in [42] offers a robust performance and is widely employed in similar medical imaging applications. Finally, the intensity range of all images was rescaled to $[0, 100]$.

3.3. Atlas Selection and ROI Extraction

The construction of an atlas library for each target image $\mathcal{T}$ to be segmented necessitates the registration of all atlases $\{\mathcal{A}_i\}_{i=1}^{n_A}$ to the particular target image. An affine transformation was employed, registering all atlas images in the target image domain space, accounting for deformations of linear nature, such as rotations, translations, shearing, scaling, etc. The same transformation was also applied to the corresponding label map $\mathcal{L}_i$ of each atlas, resulting in the atlas library $\{\mathcal{A}_i^{\mathcal{T}}, \mathcal{L}_i^{\mathcal{T}}\}_{i=1}^{N_A}$ registered to $\mathcal{T}$.

Considering the fact that the cartilage volume accounts for a very small percentage of the overall image volume, a region of interest (*ROI*) was defined for every target image, covering the entire cartilage structure and its surrounding area. A presegmentation mask was constructed by passing the registered atlas cartilage mask through a majority voting (*MV*) filter, and then expanded by a binary morphological dilation filter, yielding the *ROI* for the target image. This region corresponded to the sampling volume for the target image $\mathcal{T}$ and its corresponding atlas library $\{\mathcal{A}_i, \mathcal{L}_i\}_{i=1}^{N_A}$. This process guaranteed that the selected *ROI* enclosed the totality of cartilage tissue both in the target $\mathcal{T}$, as well as in the corresponding atlas library.

Finally, to simultaneously reduce the computational load and increase the spatial correspondence between target and atlas images, we included a final atlas selection step. Measuring the spatial misalignment in the *ROI* of every pair $\{\mathcal{T}, \mathcal{A}_i\}_{i=1}^{\mathcal{T}}$ using the mean squared difference (MSD_i^{ROI}), we only kept the first $N_{\mathcal{A}}$ atlases exhibiting the least disagreement in the metric [37].

4. Graph Constructions

In this section, we describe the node representation, the construction of aligned sequences of nodes, and the sequence libraries, which lead to the formation of the aligned image graphs used in the convolutions.

4.1. Node Representation

An important issue to properly address is how an image is transformed to a graph structure of nodes. In our setting, a node was described by a generic $5 \times 5 \times 5$ patch $\mathbf{p}_i = \mathbf{p}(\mathbf{x}_i)$ surrounding a central voxel $\mathbf{x}_i$. The image was then represented by a collection of nodes which were spatially distributed across the *ROI* volume.

Each node was described by a feature vector $\mathbf{x}_i = f_{enc}(\mathbf{p}_i) \in \mathcal{R}^{20}$ implemented via *HOG* descriptors [43], which aggregated the local information on the node patch. *HOG* descriptors constitute a staple feature descriptor in image processing and recognition. Here, we applied a modification suitable for operating on 3D data [44]. For each voxel $\mathbf{x}_i$, we extracted an *HOG* feature description by computing the gradient magnitude and direction along the $x - y - z$ axes for each constituent voxel in the node patch. The resulting values were binned to a ($q = 20$)-dimensional feature vector where each entry corresponded to the vertex of a regular icosahedron, with each bin representing the strength of the gradient along that particular direction.

Finally, each node was associated with a class indicator vector $\mathbf{y}_i = [\mathbf{y}_{i_1}, \ldots, \mathbf{y}_{i,c}] \in \mathcal{R}^c$, where $\mathbf{y}_{i,c} = 1$ if voxel $\mathbf{x}_i$ belongs in class i and 0 otherwise.

4.2. Aligned Image Graphs

The underlying principle of the multi-atlas approach is that the target image $\mathcal{T}$ and the atlas library $\mathcal{A}_i^{\mathcal{T}}$ are aligned via affine registration, thus sharing a common coordinate space. This allows the transfer of label information from atlas images towards the target one by operating upon sequences of spatially correspondent voxels. Complying with the multi-atlas setting, we applied a two-stage sampling process with the aim to construct a sequence of aligned graphs, involving the target image and its respective atlases. This sequence contained the so-called root nodes which were distinguished from the neighboring nodes introduced in the sequel.

1. *Target graph construction*: This step used a spatially stratified sampling method to generate an initial set of target voxels $\mathbf{X}_r^{\mathcal{T}} \in \mathcal{R}^{n_{\mathcal{T}} \times D}$, where D denotes the feature dimensionality. To ensure a uniform spatial covering of all classes in the target *ROI*, we performed a spatial clustering step partitioning all contained voxels into $n_r^{\mathcal{T}}$ clusters. After interpolating the cluster centers to the nearest grid point, we obtained the global

dataset $\mathbf{X}_r^{\mathcal{T}} = \{\mathbf{x}_r^{\mathcal{T}}(i), i = 1, \ldots, n_r^{\mathcal{T}}\}$, which defined a corresponding target graph of root nodes $\mathcal{G}_r^{\mathcal{T}}$. These target nodes served as reference points from which the aligned sequences were subsequently generated.

2. *Sequences of aligned data*: For each $\mathbf{x}_r^{\mathcal{T}}(i) \in \mathbf{X}_r^{\mathcal{T}}$, we defined a sequence of aligned nodes $\mathcal{S}(i)$, containing the target node $\mathbf{x}_r^{\mathcal{T}}(i)$ and its respective nodes from the atlas library, located at spatially correspondent positions:

$$\mathcal{S}(i) = \left\{\mathbf{x}_r^{\mathcal{T}}(i); \mathbf{x}_r^{\mathcal{A}_1}(i), \ldots, \mathbf{x}_r^{\mathcal{A}_{n_{\mathcal{A}}}}(i)\right\}, \quad \forall i = 1, \ldots n_{\mathcal{T}} \tag{11}$$

The entire global dataset of root nodes, containing all sampled target nodes and their associated atlas ones, was defined as the union of all those sequences

$$\mathbf{X}_r = \bigcup_{i=1}^{n_{\mathcal{T}}} \mathcal{S}(i) = \left[\mathbf{X}_r^{\mathcal{T}}, \mathbf{X}_r^{\mathcal{A}}\right] \tag{12}$$

where $\mathbf{X}_r \in \mathbb{R}(N_r \times D)$ contains a total number of $N_r = n_{\mathcal{T}} \cdot (n_{\mathcal{A}+1})$ root nodes, while $\mathbf{X}_r^{\mathcal{T}}$ and $\mathbf{X}_r^{\mathcal{A}}$ denote the datasets of root nodes sampled from the target and atlases, respectively. Accordingly, this led to a sequence of aligned graphs $\mathcal{G}_r = \{\mathcal{G}_r^{\mathcal{T}}; \mathcal{G}_r^{\mathcal{A}_1}, \ldots \mathcal{G}_r^{\mathcal{A}_{n_{\mathcal{A}}}}\}$, which is schematically shown in Figure 3. In this figure, we can distinguish two modes of pairwise relationships among root nodes that should be explored. Concretely, there are local spatial affinities across the horizontal axis between nodes belonging to a specific node sequence. On the other hand, there also exist global pairwise affinities between nodes of each image individually, as well as between nodes belonging to different images in the sequence. The latter type of search ensures that nodes of the same class located at different positions in the ROI volume and with different textural appearance are taken into consideration, thus leading to the extraction of more expressive node representations of the classes via learning.

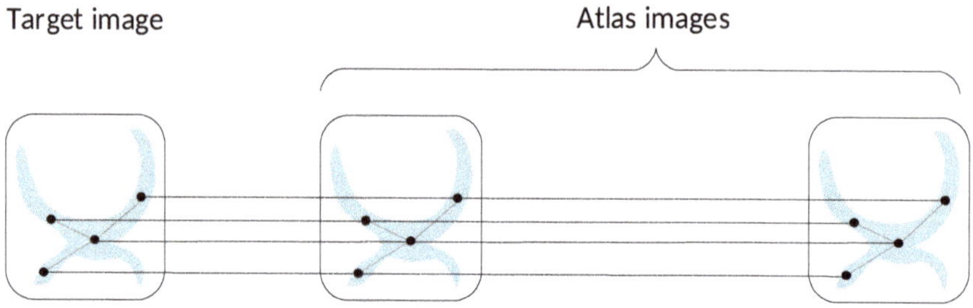

Figure 3. Schematic illustration of a sequence of aligned image graphs of root nodes, including the target graph (**left**) and the graphs of its corresponding atlases (**right**). There are local spatial affinities at aligned positions (horizontal axis), as well as global pairwise similarities between nodes located at different positions in the *ROIs*.

4.3. Sequence Libraries

It should be stressed that the cost-effective affine registration used in our method is not capable of coping with severe image deformations. Hence, it cannot provide sufficiently accurate alignment between the target and the atlases. To account for this deficiency, we expanded the domain of local search by considering neighborhoods around nodes. Specifically, for each node $\mathbf{x}_i$, we defined multihop neighborhoods at multiple scales:

$$\mathcal{R}_s(\mathbf{x}_i) = \mathcal{R}_{s-1}(\mathbf{x}_i) \cup \mathcal{R}_1(\mathcal{R}_{s-1}(\mathbf{x}_i)) \tag{13}$$

for $s = 1, \ldots, S$, where $\mathcal{R}_s(\mathbf{x}_i)$ denotes the neighborhood at scale s, and S is the number of scales used. $\mathcal{R}_0(\mathbf{x}_i) = \mathbf{x}_i$ corresponds to the basic patch $5 \times 5 \times 5$ of the node itself. $\mathcal{R}_1(\mathbf{x}_i)$

and $\mathcal{R}_2(\mathbf{x}_i)$ are the 1-hop and 2-hop neighborhoods delineated as $9 \times 9 \times 9$ and $13 \times 13 \times 13$ volumes around $\mathbf{x}_i$, respectively. In our experiments, we considered two different spatial scales ($S = 2$). Figure 4 illustrates the different node neighborhoods.

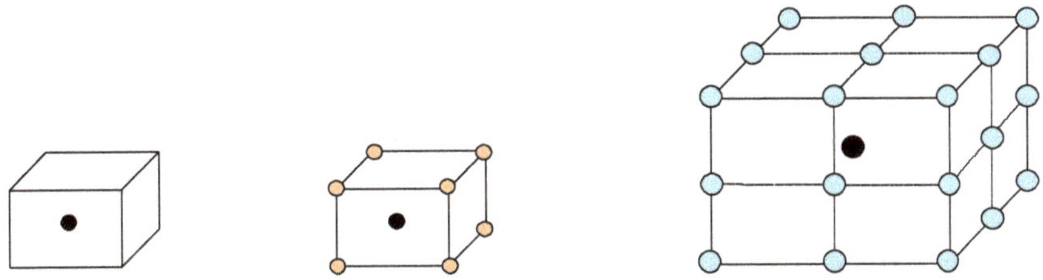

Figure 4. A generic $5 \times 5 \times 5$ patch (**left**) representing a node ($s = 0$). The corresponding 1-hop ($s = 1$, **middle**) and 2-hop neighborhoods ($s = 2$, **right**), corresponding to $9 \times 9 \times 9$ and $13 \times 13 \times 13$ hypercubes, respectively. Black dots correspond to root nodes, while colored ones stand for the neighboring nodes. All nodes are represented by $5 \times 5 \times 5$ patches.

Next, for each sequence $\mathcal{S}(i)$ $i = 1, \cdots, n_{\mathcal{T}}$, we created the corresponding sequence libraries by incorporating the local neighborhoods of all root nodes belonging to that sequence. The sequence library $\mathcal{SL}_s(i)$ at scales $s = 0, 1, \cdots, S$ was defined by

$$\mathcal{SL}_s(i) = \mathcal{R}_s(\mathbf{x}_r^{\mathcal{T}}(i)) \cup \left\{ \mathcal{R}_s(\mathbf{x}_r^{\mathcal{A}_1}(i)) \cup \ldots \cup \mathcal{R}_s(\mathbf{x}_r^{\mathcal{A}_{n_{\mathcal{A}}}}(i)) \right\} \quad (14)$$

$\mathcal{SL}_s(i)$ contains $(n_{\mathcal{A}} + 1) \cdot |\mathcal{R}_s|$ nodes, where $|\mathcal{R}_s|$ denotes the size of the spatial neighborhood at scale s. Figure 5 provides a schematic illustration of a sequence library.

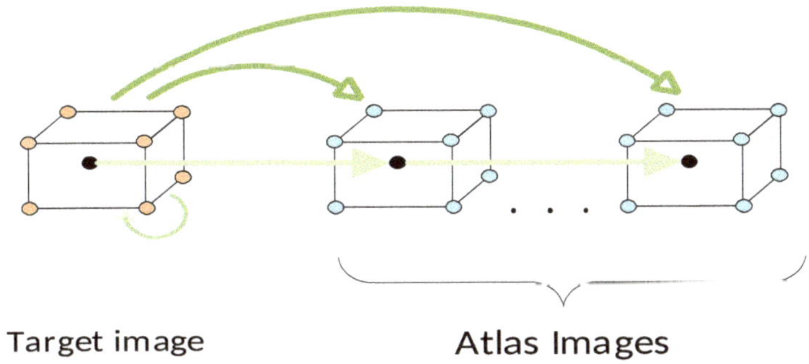

Figure 5. Schematic illustration of a sequence library for a specific scale $s = 1$, comprising the aligned neighborhoods from the target and the atlas images. Green arrows indicate the different scopes of the attention mechanism. For a particular root node, attention is paid to its own neighborhood, as well as the other neighborhoods in the sequence.

The collection of all $\mathcal{SL}_s$ forms a global dataset $\mathbf{X}_s$ of aligned neighborhoods described as follows:

$$\mathbf{X}_s = \bigcup_{i=1}^{n_{\mathcal{T}}} \mathcal{SL}_s(i) = \mathbf{X}_r \cup \mathcal{N}_s \quad (15)$$

for $s = 1, \ldots, S$. $\mathbf{X}_s$ is formed as union of the dataset $\mathbf{X}_r$ of root nodes and the dataset $\mathcal{N}_s$ comprising their neighboring nodes at scale s. Its cardinality is $|\mathbf{X}_s| = \mathcal{N}_r + |\mathcal{N}_s|$, where $\mathcal{N}_r$ is the number of root nodes, and $|\mathcal{N}_s|$ the cardinality of $\mathcal{N}_s$. Further, $\mathbf{X}_s$ corresponds to a subgraph $\mathcal{G}_s(\mathcal{V}_s, \mathcal{E}_s)$, with $|\mathcal{V}_s| = |\mathbf{X}_s|$. In this subgraph, connective edges are established in $\mathcal{E}_s$ along the horizontal axis, namely, between root nodes and the neighboring nodes across the sequence libraries. Concluding, since the neighborhoods are by definition inclusive as the scale increases, the dataset $\mathbf{X}_S$ contains the maximum number of nodes, forming the overall dataset $\widetilde{\mathbf{X}}$:

$$\widetilde{\mathbf{X}} = \mathbf{X}_S = \bigcup_{i=1}^{n_{\mathcal{T}}} \mathcal{SL}_S(i) \tag{16}$$

The corresponding graph $\mathcal{G}$ comprises an overall total number of $\mathcal{N} = |\mathcal{N}_r| + |\mathcal{N}_S|$ nodes, including the root and neighboring nodes.

5. Convolutional Units

This section elaborates on the basic convolutional units, namely, the local convolutional unit *Lconv* and the global convolutional unit *Gconv* which serve as structural elements to devise our proposed models (Figure 6).

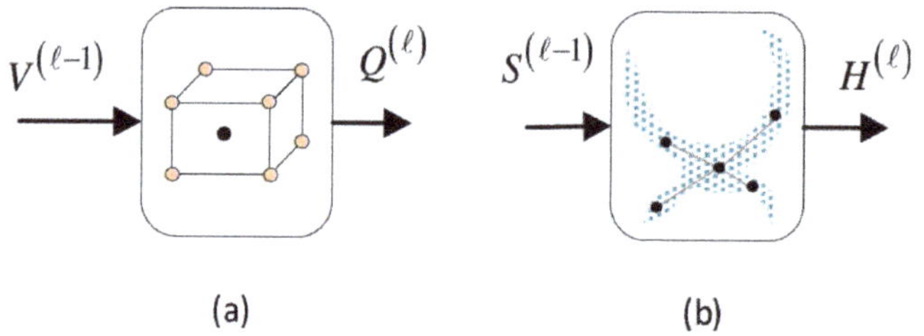

Figure 6. Outline of the convolutional units employed. (**a**) The local convolutional unit $(Lconv)$, (**b**) the global convolutional unit $(Gconv)$.

5.1. Local Convolutional Unit

The local convolutional unit undertakes the local spatial learning task, operating horizontally along the sequence libraries (SL_s) of nodes. Instead of confining ourselves to predefined and fixed weights in the graphs, we opted to apply a local attention mechanism to adaptively learn the graph structure information at each layer. Specifically, we used the attention approach suggested in the *GAT* as a means to capture the local contextual relationships among nodes in the search area.

A functional outline of *Lconv* at layer l is shown in Figure 7. It receives an input $\mathbf{V}^{(l-1)} \in \mathbb{R}^{N \times E_{in}^{(l-1)}}$ from the previous layer and provides its output $\mathbf{Q}^{(l)} \in \mathbb{R}^{N \times E_o^{(l)}}$, where $E_{in}^{(l-1)}$ and $E_o^{(l)}$ denote the dimensionalities of the input and output node features, respectively.

Figure 7 provides a detailed architecture of *Lconv*. The model involves S sub-modules, each one associated with a specific spatial scale of aggregation. The sub-module s acts upon the subgraph $\mathcal{G}_s(\mathcal{V}_s, \mathcal{E}_s), s = 0, \ldots, S$, which subsumes the sequence libraries $\mathcal{SL}_s$ of root nodes. Its input is $\mathbf{V}_s \in \mathbb{R}^{|\mathbf{X}_s| \times E_{in}^{(l-1)}}$ and after a local convolution at scale s, it provides its own output $\mathbf{Q}_s^{(l)} \in \mathbb{R}^{|\mathbf{X}_s| \times E_o^{(l)}}$. In this context, the structure of $\mathcal{G}_s$ is adapted to the local attention mechanism. Let us assume that a root node $\mathbf{x}_i$ belongs to the qth sequence library: $\mathbf{x}_i \in \mathcal{SL}_s(q), \quad q = 1, \ldots, n_{\mathcal{A}}$. Then, node $\mathbf{x}_i$ pays attention to two pools of neighboring nodes (Figure 3): (a) it aggregates relevant feature information from nodes $\mathbf{x}_j$ of its own neighborhood, $\mathbf{x}_j \in \mathcal{R}_s(\mathbf{x}_i)$ (self-neighborhood attention); (b) it aggregates features of nodes belonging to the other aligned neighborhoods in $\mathcal{SL}_s(q)$ pertaining to the atlas images: $\mathbf{x}_j \in \mathcal{SL}_s(q) \setminus \mathcal{R}_s(\mathbf{x}_i)$. For these pairs of nodes, we compute normalized attention coefficients using Equation (6). Further, pairwise affinities between nodes belonging to different sequence libraries are disregarded, i.e., $\alpha_{ij} = 0$ when $\mathbf{x}_i \in \mathcal{SL}_s(q)$ and $\mathbf{x}_j \in \mathcal{SL}_s(p)$, $p \neq q$. It should be noticed that we are primarily focused on computing comprehensive feature representations of the root nodes. Nevertheless, neighboring nodes are also updated; however, in this case, the attention is confined to the neighborhood of the root node it belongs to.

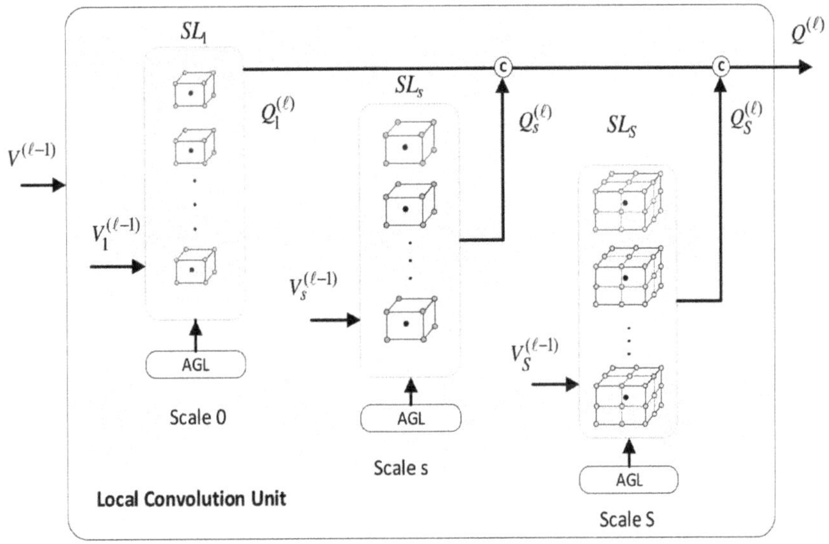

Figure 7. Detailed description of the local convolutional unit, which aggregates local contextual information from node neighborhoods at different spatial scales.

For convenience, let us consider the input node features of the form $\mathbf{V}_s^{(l-1)} = [\mathbf{v}_{s,1}^{(l-1)}, \ldots, \mathbf{v}_{s,|\mathbf{X}_s|}^{(l-1)}]$, where $\mathbf{v}_{s,i} \in \mathbb{R}^{E_{in}^{(l-1)}}$ and similarly, $\mathbf{Q}_s^{(l)} \in [\mathbf{q}_{s,1}^{(l)}, \ldots, \mathbf{q}_{s,|\mathbf{X}_s|}^{(l)}], \mathbf{q}_{s,i}^{(l)} \in \mathbb{R}^{E_o^{(l-1)}}$. The local-level convolution at scale s of a root node $\mathbf{x}_i \in \mathcal{SL}_s(q), \quad i = 1, \ldots, \mathcal{N}_r$ is obtained by:

$$\mathbf{q}_{s,i}^{(l)} = \sigma\left(\sum_{j \in \mathcal{R}_s(\mathbf{x}_i)} \alpha_{ij}^{(k)} \mathbf{W}_{s,k}^{(l)} \mathbf{v}_{s,i}^{(l-1)} + \sum_{j \in \mathcal{SL}_s(q) \setminus \mathcal{R}_s} \alpha_{ij}^{(k)} \mathbf{W}_{s,k}^{(l)} \mathbf{v}_{s,i}^{(l-1)}\right) \tag{17}$$

The first term in the above equation refers to the self-neighborhood attention, which aggregates node features from $\mathcal{R}_s(\mathbf{x}_i)$ within the same image. Moreover, the second term aggregates node information from the other aligned neighborhoods in the sequence. In an attempt to stabilize the learning process and further enhance the local feature representations, we followed a multi-head approach, whereby K independent attention mechanisms are applied. Accordingly, the node convolutions proceed as follows:

$$\mathbf{q}_{s,i}^{(l)} = \big\Vert_{k=1}^{K} \sigma\left(\sum_{j \in \mathcal{R}_s(\mathbf{x}_i)} \alpha_{ij}^{(k)} \mathbf{W}_{s,k}^{(l)} \mathbf{v}_{s,i}^{(l-1)} + \sum_{j \in \mathcal{SL}_s(q) \setminus \mathcal{R}_s} \alpha_{ij}^{(k)} \mathbf{W}_{s,k}^{(l)} \mathbf{v}_{s,i}^{(l-1)}\right) \tag{18}$$

where $\alpha_{ij}^{(k)}$ denote the attention coefficients between nodes $\mathbf{x}_i$ and $\mathbf{x}_j$ according to the kth attention head, while $\mathbf{W}_{s,k}^{(l)} \in \mathbb{R}^{E_{in}^{(l)} \times E_o^{(l-1)}}$ are the corresponding parameter weights used for node embeddings. The attention parameters are shared across all nodes in $\mathcal{G}_s(\mathcal{V}_s, \mathcal{E}_s)$ and are simultaneously learned at each layer l and for each spatial scale, individually. The outputs of the different sub-modules are finally aggregated to yield the overall output of the *Lconv* unit:

$$\mathbf{Q}^{(l)} = \mathbf{Q}_1^{(l)} \oplus \cdots \oplus \mathbf{Q}_S^{(l)} \tag{19}$$

where $\oplus$ denotes the concatenation operator.

The multilevel attention-based aggregation of valuable contextual information from sequence libraries offers some noticeable assets to our approach: (a) it acquires comprehensive node representations which assist in producing better segmentation results, (b) the graph learning circumvents the inaccuracies caused by affine registration in severe image deformations which may lead to node misclassification.

5.2. Global Convolutional Unit

The global convolutional unit conducts the global convolution task; it acts upon the subgraph $\mathcal{G}_r(\mathcal{V}_r, \mathcal{E}_r)$ which includes the sequence of root nodes $\mathcal{S}(i)$. *Gconv* aims at exploring the global contextual relationships among nodes located at different positions in the target image and the atlases (Figure 3). Accordingly, we established in $\mathcal{E}_r$ suitable pairwise connective weights according to spectral similarity $\widetilde{A}_{ij} \neq 0$ for nodes $\mathbf{x}_i \in \mathcal{S}(p)$, $\mathbf{x}_j \in \mathcal{S}(q), p \neq q$. Node pairs belonging to the same sequence are processed by *Lconv* units; hence, they are disregarded in this case.

The global convolution at layer l is acquired using the spectral convolutional principles in *GCN*:

$$\mathbf{H}^{(l)} = \sigma\left(\widetilde{\mathbf{A}}^{(l)} \mathcal{S}^{(l-1)} \mathbf{W}_g^{(l)}\right) \tag{20}$$

where $\mathcal{S}^{(l-1)} \in \mathbb{R}^{(N \times F_{in}^{(l)})}$ and $\mathbf{H}^{(l-1)} \in \mathbb{R}^{(N \times F_o^{(l)})}$ denote the input and output of *Gconv*, respectively, whereas $F_{in}^{(l-1)}, F_o^{(l)}$ are the corresponding dimensionalities. $\mathbf{W}_g^{(l)}$ is the learnable embedding matrix and $\widetilde{\mathbf{A}}^{(l)}$ is the adjacency matrix, as defined in Section 2.1.

Similar to *Lconv*, we also incorporated the *AGL* mechanism to *Gconv*, so that global affinities could be automatically captured at each layer via learning. More concretely, we applied an adaptive scheme whereby the connective weights between nodes are determined from the module's input signals [45]. The adjacency matrix elements were computed by:

$$\widetilde{\mathbf{A}} = \sigma\left[\left(\widetilde{\mathbf{H}}^{(l-1)} \mathbf{W}_\phi\right)\left(\widetilde{\mathbf{H}}^{(l-1)} \mathbf{W}_\phi\right)^T\right] + \mathbf{I}_{N \times N} \tag{21}$$

where $\widetilde{\mathcal{S}}^{(l-1)} = \mathcal{BN}(\mathcal{S}^{(l-1)})$ is obtained after applying batch-normalization to the inputs, $\sigma(\cdot)$ is the sigmoidal activation function applied on an element-wise operation, and $\mathbf{W}_\phi$ is the embedding matrix to be learned, shared across all nodes of $\mathcal{G}_r(\mathcal{V}_r, \mathcal{E}_r)$. The adaptation scheme in Equation (21) assigns greater edge values between nodes with high spectral similarity and vice-versa.

In the descriptions above, we considered the *GCN* model equipped with *AGL* as a baseline scheme. Nevertheless, in our experimental investigation, we examined several scenarios whereby the global convolution task was tackled using alternative convolutional models, including *GraphSage*, *GAT*, *GraphSAINT*, etc.

6. Proposed Convolutional Building Blocks

In this section, we present two alternative building models, namely, the cross-talk building model (*CT-BM*) and the sequential building model (*SEQ-BM*). They are distinguished according to the way the local and global convolutional units are blended across the layers. Every constituent local and global unit within the structures has its individual embedding matrix of learnable parameters. Further, it is also equipped with its own *AGL* mechanism for adaptive learning of the graphs, as described in the previous section.

6.1. Cross-Talk Building Model (CT-BM)

The *CT-BM* is shown in the outline of our approach in Figure 1. Nevertheless, a more compact form is depicted in Figure 8. The model comprises two composite layers ($l = 1, 2$), each one containing one local and one global unit. As can be seen, convolutions proceed in an alternating manner across the layers, whereby the local unit transmits its output to the next global unit, and vice versa. A distinguishing feature of this structure is that there are also skip connections and aggregators which implement cross-talk links between the local and global components. Particularly, in addition to the standard flow from one unit to the next, each unit's output is aggregated with the output of the subsequent unit.

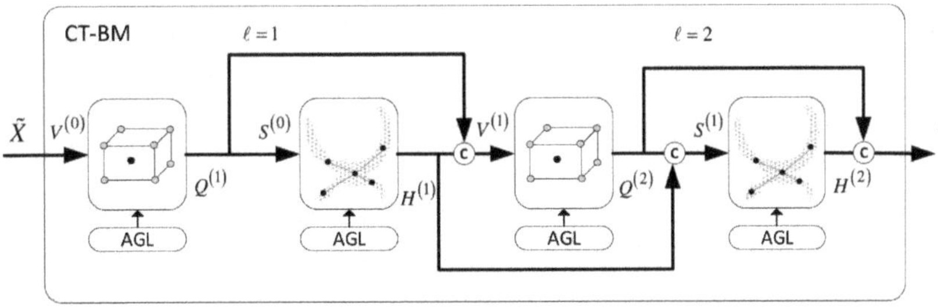

Figure 8. Illustration of the proposed cross-talk building model (*CT-BM*), where local and global convolutional units are combined following an alternating scheme.

The overall workflow of the *CT-BM* is outlined below:

1. The first local unit yields

$$\mathbf{V}^{(0)} = \mathbf{X}, \mathbf{Q}^{(1)} = Lconv(\mathbf{V}^{(0)}) \tag{22}$$

2. The local unit's output is passed to the first global unit to compute

$$\mathbf{S}^{(0)} = \mathbf{Q}^{(1)}, \mathbf{H}^{(1)} = Gconv(\mathbf{S}^{(0)}) \tag{23}$$

3. The second local unit receives an aggregated signal to provide its output,

$$\mathbf{V}^{(1)} = \mathbf{H}^{(1)} + \mathbf{Q}^{(1)}, \quad \mathbf{Q}^{(2)} = Lconv(\mathbf{V}^{(1)}) \tag{24}$$

4. The second global unit produces

$$\mathbf{S}^{(1)} = \mathbf{H}^{(1)} + \mathbf{Q}^{(2)}, \quad \mathbf{H}^{(2)} = Gconv(\mathbf{S}^{(1)}) \tag{25}$$

5. The final output of the CT-BM is the obtained by

$$\mathbf{Z} = \mathbf{H}^{(2)} + \mathbf{Q}^{(2)} \tag{26}$$

6.2. Sequential Building Model (SEQ-BM)

The architecture of the *SEQ-BM* is illustrated in Figure 9. This model also contains two local and two global units. Contrary to the *CT-BM*, convolutions are conducted in the *SEQ-BM* sequentially. Concretely, the local learning task is first completed using the first two local convolutional units. The outputs of this stage are then transmitted to the subsequent stage which accomplishes the global learning task, using the two global convolutional units. The overall output of the *SEQ-BM* is formed by aggregating the resulting local and global features of the two stages.

The workflow of the *SEQ-BM* is outlined as follows:

1. The local learning task is described by

$$\mathbf{V}^{(0)} = \mathbf{X}, \mathbf{Q}^{(1)} = Lconv(\mathbf{V}^{(0)}) \tag{27}$$
$$\mathbf{V}^{(1)} = \mathbf{Q}^{(1)}, \mathbf{Q}^{(2)} = Lconv(\mathbf{V}^{(1)}) \tag{28}$$

2. The global learning task is described by

$$\mathbf{S}^{(0)} = \mathbf{Q}^{(2)}, \mathbf{H}^{(1)} = Gconv(\mathbf{S}^{(0)}) \tag{29}$$

$$\mathbf{S}^{(1)} = \mathbf{H}^{(1)}, \mathbf{H}^{(2)} = Gconv(\mathbf{S}^{(1)}) \tag{30}$$

3. The final output of the SEQ-BM is obtained by

$$\mathbf{Z} = \mathbf{Q}^{(2)} + \mathbf{H}^{(2)} = \mathbf{Z}_{loc} + \mathbf{Z}_{glo} \tag{31}$$

The alternating blending of the *CT-BM* provides a more effective integration between local and global features at each layer, individually, as compared to the sequential combination in *SEQ-BM*. This observation is attested experimentally as shown in Section 10.

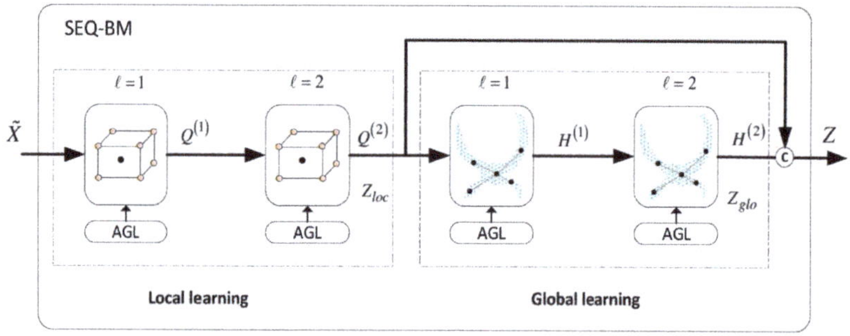

Figure 9. Illustration of the proposed sequential building model (*SEQ-BM)*, where the local and global learning tasks are carried out sequentially.

7. Proposed Dense Convolutional Networks

In this section, we present two variants of our main model, the *DMA-GCN* network. The motivation behind this design is based on an attempt to further expand the structures *CT-BM* and *SEQ-BM* by including multiple layers of convolutions, to face the gradient vanishing effect, where the gradients diminish, thus hindering effective learning or even worsening the results. This is the reason why in the above building blocks, we are restricted to two-layered local–global convolutions.

To tackle this problem, we resorted to the recent advancements in deep GCNs [36] and developed the *DMA-GCN* model with a densely connected convolutional architecture utilizing residual skip connections, as shown in Figure 10. As can be seen, the model consists of several blocks arranged across M layers of block convolutions. These blocks are implemented by either *CT-BM* or *SEQ-BM* described in Section 5, which leads to two different alternative configurations, the *DMA-GCN(CT-BM)* and *DMA-GCN(SEQ-BM)*, respectively. Let $\mathbf{Z}_{in}^{(i)} \in \mathbb{R}^{(N \times P_{in}^{(i)})}$ and $\mathbf{Z}_{o}^{(i)} \in \mathbb{R}^{(N \times P_{o}^{(i)})}$ denote the input and output of the ith block, $i = 1, \ldots, M$, with $P_{in}^{(i)}, P_{out}^{(i)}$ being the feature dimensionalities, respectively. The properties of the suggested *DMA-GCN* are discussed in the following:

1. The skip connections interconnect the blocks across the layers. Concretely, each block receives as input the outputs of blocks from all preceding layers:

$$\mathbf{Z}_{in}^{(i)} = \mathbf{Z}_{o}^{(1)} \oplus \mathbf{Z}_{o}^{(2)} \oplus \cdots \oplus \mathbf{Z}_{o}^{(i-1)} \tag{32}$$

for $i = 1, \ldots, M$, where $\oplus$ denotes the concatenation operator. This allows the generation of deeper *GCN* structures which can acquire more expressive node features. Overall, the *DMA-GCN* involves $4M$ convolutional units. Within each block, two layers of local–global convolutions are internally performed; the resulting outputs are then integrated along the block layers to provide the final output:

$$\mathbf{Z}_o = \mathbf{Z}_o^{(1)} \oplus \mathbf{Z}_o^{(2)} \oplus \ldots \oplus \mathbf{Z}_o^{(M)} \tag{33}$$

2. The other beneficial effect of skip connections is that they allow the final output to have direct access to the outputs of all blocks in the dense network. This assures a better reverse flow of information and facilitates the effective learning of parameters pertaining to the blocks. Since block operations are confined to two-layered local–global convolutions, overall, we can circumvent the gradient vanishing problem.
3. In order to preserve the parametric complexity at a reasonable level, similar to [36], we define the feature dimensions of each block in *DMA-GCN* to be the same:

$$P_o^{(1)} = P_o^{(2)} = \cdots = P_o^{(M)} = d \tag{34}$$

The node feature growth rate caused by the aggregators can be defined as $P_{in}^{(i)} = (i-1) \cdot d, i = 1, \cdots, M$. The input dimensions grow linearly as we proceed to deeper block layers, with the last block showing the largest increase $P_{in}^{(M)} = (M-1) \cdot d$. To prevent feature dimensionalities from receiving too large values, we considered initially a *DMA-GCN* model with $M = 4$ blocks. The particular number of blocks in the above range was then decided after experimental validation (Section 10).
4. Every block in *DMA-GCN* is supported with its corresponding *AGL* process to automatically learn the graph connective affinities at each layer. This is accomplished by applying an attention-based mechanism for local convolutional units (Section 5.1) and an adaptive construction of adjacency matrices from inputs node features (Section 5.2).

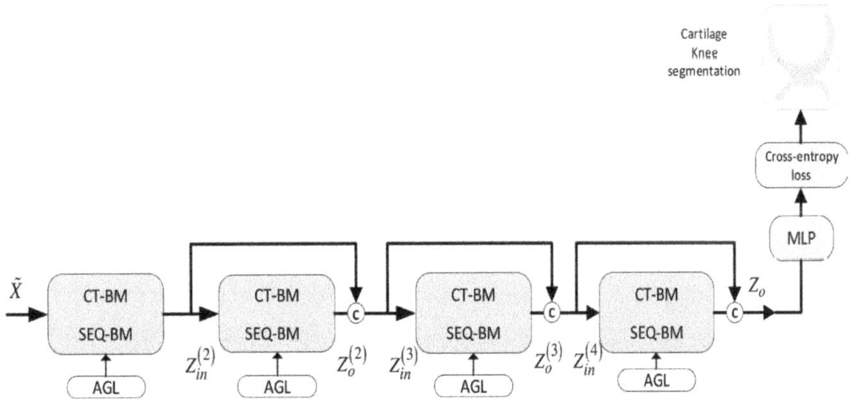

Figure 10. Description of the proposed *DMA-GCN* model, with densely connected block convolutional structure and residual skip connections.

The output of the *DMA-GCN* is fed to a two-Layer *MLP* unit to obtain the label estimates

$$\widehat{\mathbf{Z}_o} = \text{softmax}(ReLU(\mathbf{W}_1\mathbf{Z})\mathbf{W}_2) \tag{35}$$

We adopted the cross-entropy error to penalize the differences between the model's output $\widehat{\mathbf{Z}}_o$ and the corresponding node labels

$$\mathcal{L} = -\sum_{l \in \mathcal{Y}_L} \sum_{c=1}^{C} \mathbf{Y}_{lc} \mathbf{Z}_{lc} \tag{36}$$

where $\mathcal{Y}_L$ denotes the subset of labeled nodes. The *DMA-GCN* network was trained under either the transductive or the inductive learning methods, as discussed in the following section.

8. Network Learning

In our setting, we considered transductive learning (*SSL*) as the basic learning scheme for training the *DMA-GCN* models. In this case, both unlabeled data from the target image to be segmented $\mathcal{T}$ and the labeled data from the atlases $\mathcal{LB}(\mathcal{T})$ were used for the construction of the model. Nevertheless, in the experiments, we also investigated the inductive (supervised) learning scenario, whereby training was conducted by solely using labeled data from the atlases.

8.1. Transductive Learning (SSL)

The *SSL* scheme was adapted to the context of image segmentation task elaborated here. In regard to the data used in the learning, *SSL* can be carried out along two different modes of operation, namely, mini-batch learning and full-batch learning. Next, we detail mini-batch learning and then conclude with full-batch learning, which is a special case of the former one.

Mini-batch learning was implemented by following a three-stage procedure. In stage 1, an initial model was learned and used to label an initial batch of data from $\mathcal{T}$. Stage 2 was an iterative process, where out-of-sample batches were sequentially sampled from $\mathcal{T}$ and labeled via refreshing learning. Finally, stage 3 labeled the remaining voxels of the target image using a majority voting scheme.

Stage 1: Learning. In this stage, $(t = 0)$, we started by sampling an initial unlabeled batch $\mathbf{X}_r^{\mathcal{T}}(0)$ from the target image. Then, we used the different steps detailed in Section 4 to construct the corresponding graph of nodes. (a) Given $\mathbf{X}_r^{\mathcal{T}}(0)$, we created the corresponding aligned sequences (Section 4.3), giving rise to the dataset of root nodes $\mathbf{X}_r(0) = \left[\mathbf{X}_r^{\mathcal{T}}(0), \mathbf{X}_r^{\mathcal{A}}(0)\right]$. (b) Next, we incorporated neighborhood information by generating sequence libraries at multiple scales (Section 4.3), leading to the datasets $\mathbf{X}_s(0) = \mathbf{X}_r(0) \cup \mathcal{N}_s(0), s = 1, \ldots S$, where $\mathcal{N}_s(0)$ denotes the neighboring nodes. (c) Finally, we considered the overall dataset $\widehat{\mathbf{X}}(0) = \mathbf{X}_S(0)$ that corresponded to the graph $\mathcal{G}_0(\mathcal{V}_0, \mathcal{E}_0)$ containing the root and neighboring nodes at that stage.

The next step was to perform convolutional learning on graph $\mathcal{G}_0(\mathcal{V}_0, \mathcal{E}_0)$ using *DMA-GCN* models. Upon completion of the training process, we accomplished the labeling of $\mathbf{X}_r^{\mathcal{T}}(0)$,

$$l\left(\mathbf{X}_r^{\mathcal{T}}(0)\right) = \mathcal{F}_{(0)}\left(\widetilde{\mathbf{X}}(0), \mathcal{W}(0)\right) \tag{37}$$

$\mathcal{F}_{(0)}$ denotes the model's functional mapping, $l(\cdot)$ is the labeling function of the target nodes, and $\mathcal{W}(0)$ stands for the network's weights, including the learnable parameters of embedding matrices and attention coefficients, across all layers of the *DMA-GCN*.

Stage 2: iterative learning. This stage followed an iterative procedure, $t = 1, \cdots, T$, whereby at each iteration, out-of-sample batches of yet unlabeled nodes were sampled from $\mathcal{T}$, $\mathbf{X}_o^{\mathcal{T}}(t)$ of size n_o. Considering the nodes in $\mathbf{X}_o^{\mathcal{T}}$ as root nodes, we then applied steps (a)–(c) of the previous stage, to obtain the datasets $\mathbf{X}_{o,s} = \mathbf{X}_o(0) \cup \mathcal{N}_{o,s}(0), \quad s = 0, \ldots, S$, and the overall set $\widetilde{\mathbf{X}}_o(t) = \mathbf{X}_{o,S}(t)$, which corresponded to a graph of out-of-sample nodes. In the following, data $\widetilde{\mathbf{X}}_o(t)$ were fed to the pretrained model from stage 1, $l(\mathbf{X}_o^{\mathcal{T}}) = \mathcal{F}_{(t)}(\widetilde{\mathbf{X}}_o(t), \mathcal{W}(t))$. The model was initialized as $\mathcal{W}(t) = \mathcal{W}_0$ to preserve previously acquired knowledge. Further, it was subject to several epochs of refreshing convolutional learning, with the aim of adapting to the newly presented data. The above sequential process terminated at $t = T$ when all target nodes were labeled.

Stage 3: labeling of remaining voxels. This was the final stage of target image segmentation, entailing the labeling of target voxels not considered during the previous learning stages. Given that nodes were the central voxels of a generic $5 \times 5 \times 5$ patches, there were multiple remaining voxels scattered within $3 \times 3 \times 3$ volumes. Labeling of these voxels was accomplished by a voting scheme. Specifically, for each $\mathbf{x}_r$, the voting function accounted for both the spectral and the spatial distance from its surrounding labeled vertices:

$$l(\mathbf{x}_r) = \sum_i w_{r,i} l(\mathbf{x}_i) \tag{38}$$

$$w_{r,i} = w_{r,i}^{spec} \times w_{r,i}^{spat} \tag{39}$$

where $w_{r,i}^{spec}$ and $w_{r,i}^{spat}$ are normalized weighting coefficients denoting the spectral and spatial proximity, measured by the l_2 norm (Euclidean distance) and l_1 norm (Manhattan distance), respectively.

Full-batch learning is a special case of the above mini-batch learning. In that case, the dataset $\mathbf{X}_r^{\mathcal{T}}(0)$ is a large body of data, comprising all possible target nodes contained in the target *ROI*. Under this circumstance, the iterative stage 2 is disregarded. Full-batch learning is completed after convolutional learning (stage 1), followed by the labeling of the rest of target voxels (stage 3).

8.2. Inductive Learning

Under this setting, the target data remain unseen during the entire training phase. Adapting to the multi-atlas scenario, we devised a supervised learning scheme according to the following steps. (a) For each target image $\mathcal{T}$, we selected the most similar labeled image $\widehat{\mathcal{T}} = \mathcal{NN}(\mathcal{T})$ from atlases, where $\mathcal{NN}(\cdot)$ is a spectral similarity function used to identify the nearest neighbors of $\mathcal{T}$. (b) The image $\widehat{\mathcal{T}}$ along with its corresponding atlas library $\mathcal{LB}(\widehat{\mathcal{T}})$ were used to learn a supervised model $\mathcal{F}_{ind}(\widehat{\mathcal{T}})$ by applying exactly stage 1 of the previous subsection. (c) Finally, the target image $\mathcal{T}$ was labeled by means of $\mathcal{F}_{ind}(\widetilde{\mathcal{T}})$. As opposed to the *SSL* scenario, the critical difference was that the developed model relied solely on labeled data, disregarding target image information.

9. Experimental Setup

9.1. Evaluation Metrics

The overall segmentation accuracy achieved by the proposed methods was evaluated using the following three, standard volumetric measures: the Dice similarity coefficient ($\mathcal{DSC}$), the volumetric difference ($\mathcal{VD}$), and the volume overlap error ($\mathcal{VOE}$). Denoting $\mathcal{Y}$ the ground truth labels and $\widehat{\mathcal{Y}}$ the estimated ones, the above measures are defined as:

$$\mathcal{DSC} = 100\frac{|\mathcal{Y} \cap \widehat{\mathcal{Y}}|}{|\mathcal{Y}| + |\widehat{\mathcal{Y}}|} \tag{40}$$

$$\mathcal{VOE} = 100\left(1 - \frac{\mathcal{DSC}}{200 - DSC}\right) \tag{41}$$

$$\mathcal{VD} = 100\frac{|\widehat{\mathcal{Y}}| - |\mathcal{Y}|}{|\mathcal{Y}|} \tag{42}$$

Taking into account that the large majority of voxels correspond to either the background class or the two bone classes, we opted to also include the *precision* and *recall* classification measures, to better evaluate the segmentation performance on each individual structure. All measures correspond exclusively to the image content delineated by the respective *ROI* of each evaluated MRI.

9.2. Hyperparameter Setting and Validation

The overall performance of the proposed method depends on a multitude of preset parameters, the most prominent of which are the number of atlases $N_{\mathcal{A}}$ comprising the atlas library $\{\mathcal{A}, L\}_i, i = 1, \ldots, N_{\mathcal{A}}$, the number of heads K utilized in the multi-head attention mechanism, and the number of scales S corresponding to the different neighborhood scales. The optimal values of the above hyperparameters, as well as the performance of the *DMA*-

GCN(SEQ) and *DMA-GCN(CT)* segmentation methods, were evaluated through a 5-fold cross-validation.

9.3. Experimental Test Cases

The proposed methodology comprises several components affecting its overall capacity and performance. We hereby present a series of experimental test cases, aiming to shed light on those effects.

1. *Local vs. global learning*: In this scenario, we aimed to observe the effect of performing local-level learning in addition to global learning. The goal here was to determine the potential boost in performance facilitated by the inclusion of the attention mechanism in our models.
2. *Transductive vs. inductive learning*: The goal here was to ascertain whether the increased cost accompanying the transductive learning scheme could be justified in terms of performance, as compared to the less computationally demanding inductive learning.
3. *Sparse dense adjacency matrix*: Here, we examined the effect of progressively sparsifying the adjacency matrix $\widetilde{\mathbf{A}}^{(l)}$ at each layer on the overall performance. We examined the following cases: (1) the default case with a dense $\widetilde{\mathbf{A}}^{(l)}$ and (2) thresholding $\widetilde{\mathbf{A}}^{(l)}$ so that each node was allowed connections to $5, 10$, or 20 spectrally adjacent ones.
4. *Global convolution models*: Finally, we tested the effect of varying the design of the global components by examining some prominent architectures, namely, *GCN, ClusterGCN, GraphSAINT, and GraphSAGE*

9.4. Competing Cartilage Segmentation Methods

The efficacy of our proposed method was evaluated against several published works dealing with the problem of automatic knee cartilage segmentation.

9.4.1. Patch-Based Methods

The patch-based sparse coding (PB_{SC}) [8] and patch-based nonlocal-means (PB_{NLM}) [7] methods are two state-of-the art approaches in medical image segmentation. For consistency reasons, similar to the *DMA-GCN*, we set the patch size for both these methods to $(5 \times 5 \times 5)$ and the corresponding search volume size to $(13 \times 13 \times 13)$. The remaining parameters were taken as described in their respective works.

9.4.2. Deep Learning Methods

Here, we opted to evaluate the *DMA-GCN* against some state-of-the-art deep learning architectures that were successfully applied in the field of medical image segmentation.

SegNet [12]: A convolutional encoder–decoder architecture, utilizing the state-of-the-art *VGG16* [46] network that is suitable for pixelwise classification. Since we are dealing with 3D data, we split each input image patch into its constituent planes ($-xy, -xz, -yz$) and fed those to the network.

DenseVoxNet [13]: A convolutional network proposed for cardiovascular MRI segmentation. It comprises a downsampling and upsampling sub-component, utilizing skip connections from each layer to its subsequent ones, enforcing a richer information flow across the layers. In our experiments, the model was trained using the same initialization scheme for the parameters' values as described in the original paper.

VoxResNet [14]: A deep residual network comprising a series of stacked residual modules, each one performing batch-normalization and convolution, also containing skip connections from each module's input to its respective output.

KCB-Net [16]: A recently proposed network that performs cartilage and bone segmentation from volumetric images, by utilizing a modular architecture, where initially, each one of the three sub-components is trained to process a separate plane (sagittal, coronal, axial), followed by a 3D component with the task of aggregating the respective outputs into a single overall segmentation map.

CAN3D [47]: This network utilizes a successively dilated convolution kernel aiming to aggregate multi-scale information by performing feature extraction within an increasingly dilating receptive field, facilitating the final voxelwise classification in full resolution. Additionally, the loss function employed at the final layer consists of a combination of Dice similarity coefficient $\mathcal{DSC}$ and $\mathcal{DSF}$, a variant of the standard $\mathcal{DSC}$ used in evaluating segmentation results.

Point-Net [15]: *Point-Net* is a recently proposed architecture specifically geared towards point-cloud classification and segmentation. As an initial preprocessing step, it incorporates a spatial transformer network (*STN*) [48] that renders the input invariant to permutations and is used to produce a global feature for the whole point cloud. That global feature is appended on the output of a standard multilayer perceptron (*MLP*) that operates on the initial point-cloud features, and the resulting aggregated features are passed through another *MLP* in order to provide the final segmentation map.

9.4.3. Graph-Based Deep Learning Methods

In regard to graph-based convolution models, we compared our *DMA-GCN* approach to a series of baseline *GCN* architectures to carry out the global learning task. In addition, we considered in the comparisons the multilevel *GCN* with automatic graph learning (*MGCN-AGL*) method [30], used for the classification of hyperspectral remote sensing images. The *MGCN-AGL* approach takes a form similar to the one of the *SEQ-BM* in Figure 9 to combine the local learning via a *GAT*-based attention mechanism and global learning implemented by a *GCN*. A salient feature of this method is that the global contextual affinities are reconstructed based on the node representations obtained after completion of the local learning stage.

9.5. Implementation Details

All models presented in this study were developed using the PyTorch Geometric library (https://github.com/pyg-team/pytorch_geometric, accessed on 1 May 2023), specifically built upon PyTorch (https://pytorch.org) to handle graph neural networks. For the initial registration step, we used the elastix toolkit (https://github.com/SuperElastix/elastix, accessed on 1 May 2023). The code for all models proposed in this study can be found at (https://gitlab.com/christos_chadoulos/graph-neural-networks-for-medical-image-segmentation, accessed on 10 March 2024).

Regarding the network and optimization parameters used in our study, we opted for the following choices: after some initial experimentation, the parameter d controlling the node feature growth rate (Section 7) was set to $d = 128$, resulting in the input feature dimensionality progression $128 \rightarrow 256 \rightarrow 384 \rightarrow 512$ for $M = 4$ dense layers. All models were trained for 500 epochs using the Adam optimizer, with an early stopping criterion halting the training process either when no further improvement was detected on the validation error in the span of 50 epochs, or when the validation error steadily increased for more than 10 consecutive epochs.

10. Experimental Results

10.1. Parameter Sensitivity Analysis

In this section, we examine the effect of critical hyperparameters in the performance of the models under examination. The numbers and figures presented for each hyperparameter correspond to results obtained while the remaining ones assumed their optimal determined value.

10.1.1. Number of Selected Atlases $N_{\mathcal{A}}$

The number of selected atlases is a crucial parameter for all methods adopting the multi-atlas framework. Figure 11 shows the effect on the performance of *DMA-GCN(CT)* and *DMA-GCN(SEQ)* by sampling the following values $\mathcal{N}_{\mathcal{A}} = 5, \ldots, 20$.

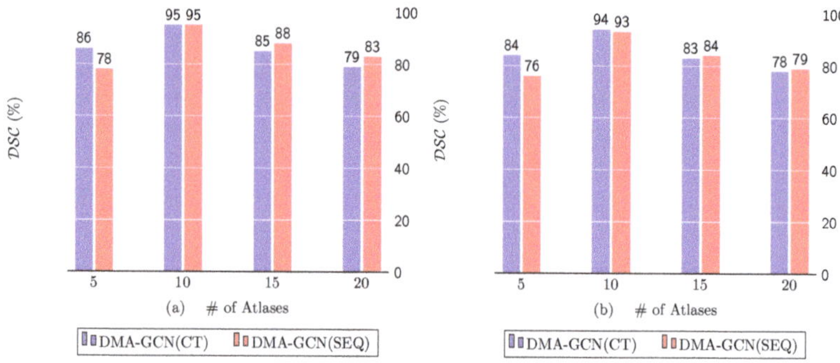

Figure 11. Cartilage $\mathcal{DSC}(\%)$ score vs. number of atlases. (**a**) Femoral cart, (**b**) tibial cart.

For both methods, the number of atlases has a similar effect on the overall performance. The highest score in each case is achieved for $\mathcal{N}_{\mathcal{A}} = 10$ atlases and slowly diminishes as that number grows. Constructing the graph by sampling voxels from a small pool of atlases increases the bias of the model, thus failing to capture the underlying structure of the data. Increasing that number allows the image graphs to include a greater percentage of nodes with dissimilar feature descriptions, which enhances the expressive power of features. Accordingly, a moderate number of aligned atlases seems to provide the best overall rates, as it achieves a reasonable balance between bias and variance.

10.1.2. Number of Attention Heads

The number of attention heads is arguably one of the most influential parameters for the local units of our models. Figure 12 demonstrates the effect on performance for $K = 0, 4, 8, 12$. The trivial case of $K = 0$ corresponds to the case where the local convolutional units are disregarded, i.e., the convolution task is undertaken solely by the global units.

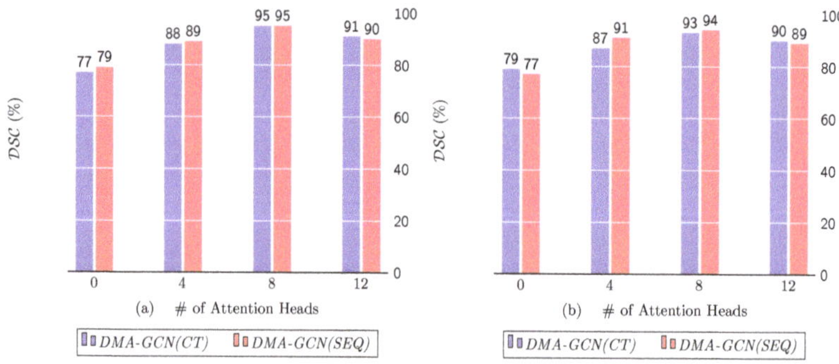

Figure 12. Cartilage $\mathcal{DSC}(\%)$ score vs. number of attention heads. (**a**) Femoral cart, (**b**) tibial cart.

The best performance is achieved for $K = 8$ for either *DMA-GCN(CT)* or *DMA-GCN(SEQ)*. Most importantly, in both charts, we can notice a sharp drop in performance when $K = 0$. This indicates that discarding the local convolutional units significantly aggravates the overall efficiency of the *DMA-GCN*. Particularly, in that case, the model disregards the local contextual information contained in node libraries, while the node features are formed by applying graph convolutions at a global level exclusively.

It can also be noted that independent embeddings of the attention mechanism provide different representations of the local pairwise affinities between nodes, which facilitates a better aggregation of the local information. Nevertheless, beyond a threshold value, the results deteriorate, most likely due to overfitting.

10.1.3. Sparsity of Adjacency Matrix $\widetilde{\mathbf{A}}$

The adjacency matrix $\widetilde{\mathbf{A}}$ is the backbone of graph neural networks in general, encoding the graph structure and node connectivity. As mentioned in [30], a densely connected $\widetilde{\mathbf{A}}$ may have a negative impact on the overall segmentation performance. To this end, we evaluated a number of thresholds that served as cut-off points, discarding edges that were not sufficiently strong. A small threshold leaves most edges in the graph intact, while a larger one creates a sparser $\widetilde{\mathbf{A}}$ by preserving only the most significant edges. Figure 13 summarizes the effects of thresholding $\widetilde{\mathbf{A}}$ at different sparsity levels.

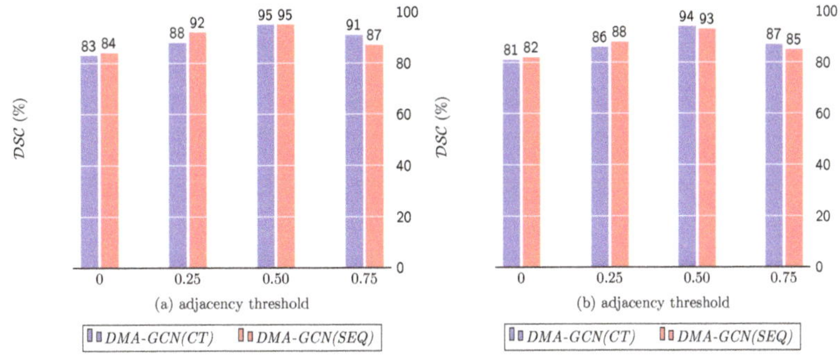

Figure 13. Cartilage $\mathcal{DSC}(\%)$ score vs. adjacency threshold. (**a**) Femoral cart, (**b**) tibial cart.

As can be seen, large sparsity values considerably reduce the segmentation accuracy, which suggests that preserving only the strongest graph edges decreases the aggregation range from neighboring nodes, and hence the expressive power of the resulting models. On the other hand, similar deficiencies are incurred for low sparsity values with dense matrices $\widetilde{\mathbf{A}}$. In that case, the neighborhood's range is unduly expanded, thus allowing aggregations between nodes with weak spectral similarity. A moderate sparsity of the adjacency matrices corresponding to a threshold value of 0.50 attains the best results for both *DMA-GCN(CT)* and *DMA-GCN(SEQ)*.

10.1.4. Number of Scales

The number of scales considered in conjunction with the local attention mechanism plays an important role in the overall performance of the DMA-GCN. It defines the size of spatial neighborhoods considered in local convolutional units, which greatly affects the resulting node feature representations. Figure 14 shows the segmentation rates for different values of scales $s = 0, 1, 2$. As can be seen, for both *DMA-GCN(CT)* and the *DMA-GCN(SEQ)* models, the incorporation of additional neighborhoods of progressively larger scales improves the accuracy results, consistently.

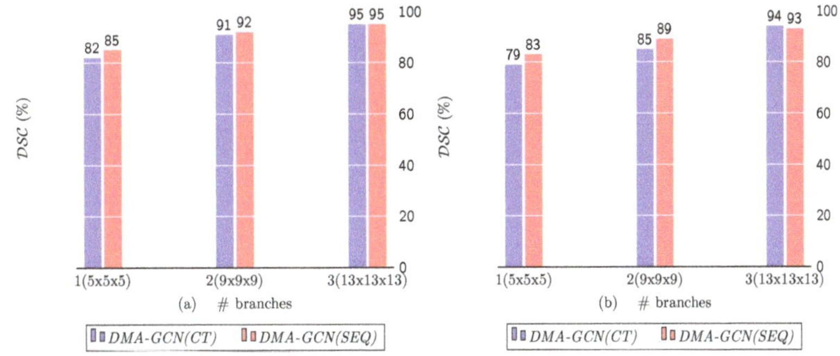

Figure 14. Cartilage $\mathcal{DSC}(\%)$ score vs. number of scales. (**a**) Femoral cart, (**b**) tibial cart.

In the trivial case of $s = 0$, each node aggregates local information by paying attention solely to its aligned root nodes. Due to the restricted attention, we were led to weak local representations of nodes, and hence degraded overall performance for the models (left columns). Incorporating the one-hop neighborhoods ($s = 1$), we expanded the range of the attention mechanism, which provided more enriched node features. This resulted in significantly better results compared to the previous case (middle columns). The above trend was further retained by including the two-hop neighborhoods of nodes ($s = 2$), where we could notice an even greater improvement of results (right columns).

10.2. Number of Dense Layers

In this section, we examine the effect of the number M of dense layers used in the *DMA-GCN* models (Section 7). It defines the depth of the networks and thus directly impacts the size, as well as the representational capabilities of the respective models. Figure 15 shows the results obtained by progressively using up to four dense layers. It should be noticed that the single layer results refer to the case where *DMA-GCN(CT)* and *DMA-GCN(SEQ)* coincide with their constituent block models, i.e., the *CT-BM* and *SEQ-BM*, respectively.

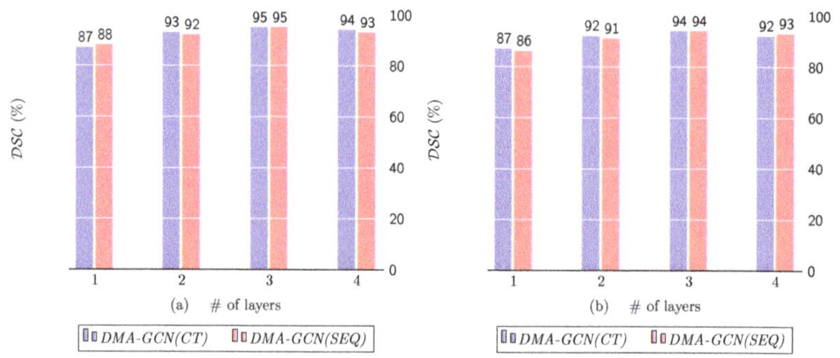

Figure 15. Cartilage $\mathcal{DSC}(\%)$ score vs. number of dense layers. (**a**) Femoral cart, (**b**) tibial cart.

Figure 15a clearly demonstrates an upwards trend in the obtained rates as the number of dense layers increases. The lowest performance is unsurprisingly achieved for the shallow network of a single layer (*CT-BM*). The best results are achieved for $M = 3$, while the inclusion of an additional layer diminishes slightly the accuracy, an effect most likely attributed to overfitting. A similar pattern of improvements can also be observed for the case of *DMA-GCN(SEQ)* model. Concluding, the proposed densely connected block architectures lead to deeper *GCN* networks with multiple local–global convolutional layers, which can acquire more comprehensive node features, and thus offer better results.

10.3. Mini-Batch vs. Full-Batch Training

The goal of this section is twofold. First, we aim to compare the mini-batch against full-batch learning schemes (Section 8). Secondly, we examine the effect of the batch size on the performance of the mini-batch learning. Figure 16a,b presents the $\mathcal{DSC}$ metrics of the models *DMA-GCN(CT)* and *DMA-GCN(SEQ)*, respectively, for varying batch sizes. The first four columns refer to the mini-batch learning, while the rightmost column represents the full-batch scenario.

Both figures share a similar pattern of results. Noticeably, in the mini-batch learning, the best results ($\mathcal{DSC} = 95\%$) were achieved for a batch size of 128. This value referred to the size of the initial batch ($\mathbf{X}_0^{\mathcal{T}}$) as well as the out-of-sample batches extracted from the target $\mathcal{T}$. Given these batches, we then proceeded to the formation of the sequences of aligned nodes and the sequence libraries at multiple scales, to generate the corresponding graphs of nodes used for convolutional learning. Finally, the full-batch learning provided a significantly inferior performance ($\mathcal{DSC} = 81\%$) compared to the mini-batch scenario.

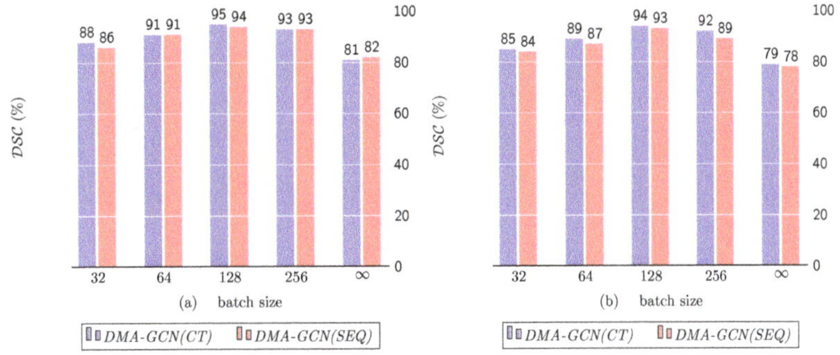

Figure 16. Cartilage $\mathcal{DSC}(\%)$ score vs. batch size. (**a**) Femoral cart, (**b**) tibial cart.

10.4. Global Module Architecture

In this section, we examine the effectiveness of some of the popular graph-based networks in our approach. Concretely, in the context of *DMA-GCN* structures, we considered several model combinations, whereby the *GAT*-based attention and local information aggregation was used to carry out the local learning task, while the global task was undertaken by the *GCN*, *SAGE*, *SAINT* and *ClustGCN*, respectively.

Figure 17 shows the $\mathcal{DSC}$ measures for both *DMA-GCN(CT)* and *DMA-GCN(SEQ)* models. Table 1 also provides more detailed results on this issue. As can be seen, for both models, the utilization of *GraphSAGE* clearly provided the best performance, possibly due to its more sophisticated node sampling method. Nevertheless, all the alternatives offered consistently good results.

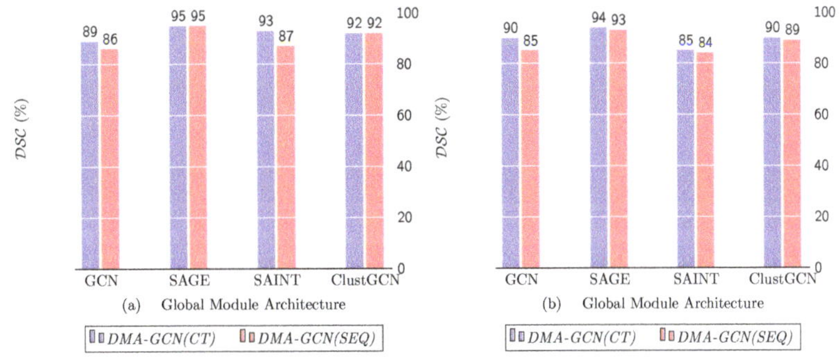

Figure 17. Cartilage $\mathcal{DSC}(\%)$ score vs. global module architecture. (**a**) Femoral cart, (**b**) tibial cart.

Table 1. Summary of segmentation performance measures (means ± stds) of the two cartilage classes of our proposed method *DMA-GCN (CT/SEQ)* by varying the global component of the sub-modules (*GCN, SAINT, SAGE, ClustGCN*). Best results for each category (*CT* vs *SEQ*) with respect to *DSC* index are highlighted.

			Femoral Cartilage					Tibial Cartilage				
Module	**CT**	**SEQ**	**Recall**	**Precision**	$\mathcal{DSC}$	$\mathcal{VOE}$	$\mathcal{VD}$	**Recall**	**Precision**	$\mathcal{DSC}$	$\mathcal{VOE}$	$\mathcal{VD}$
GAT-GCN	✓		88.26% (±0.076)	88.91% (±0.041)	**89.23%** (±0.074)	22.87% (±0.058)	7.05% (±0.048)	87.02% (±0.032)	85.36% (±0.022)	**90.12%** (±0.038)	25.36% (±0.078)	7.97% (±0.055)
		✓	87.13% (±0.077)	88.12% (±0.052)	86.49% (±0.061)	23.06% (±0.082)	7.13% (±0.029)	86.88% (±0.021)	85.03% (±0.035)	84.79% (±0.021)	25.76% (±0.041)	8.01% (±0.044)

Table 1. *Cont.*

Module	CT	SEQ	Femoral Cartilage					Tibial Cartilage				
			Recall	Precision	$\mathcal{DSC}$	$\mathcal{VOE}$	$\mathcal{VD}$	Recall	Precision	$\mathcal{DSC}$	$\mathcal{VOE}$	$\mathcal{VD}$
GAT-SAGE	✓		96.17% (±0.021)	95.81% (±0.019)	**95.71%** (±0.039)	13.17% (±0.055)	3.94% (±0.091)	95.31% (±0.029)	94.78% (±0.045)	**94.02%** (±0.036)	17.98% (±0.047)	4.99% (±0.022)
		✓	96.13% (±0.061)	95.21% (±0.071)	95.44% (±0.065)	13.21% (±0.059)	3.98% (±0.031)	95.19% (±0.025)	94.41% (±0.045)	93.87% (±0.023)	18.65% (±0.038)	5.03% (±0.044)
GAT-SAINT	✓		92.91% (±0.051)	93.43% (±0.039)	**92.87%** (±0.072)	14.02% (±0.082)	5.21% (±0.031)	88.81% (±0.023)	86.05% (±0.039)	**88.03%** (±0.057)	25.39% (±0.062)	7.87% (±0.044)
		✓	87.13% (±0.066)	88.12% (±0.045)	86.49% (±0.061)	23.06% (±0.071)	7.13% (±0.049)	87.92% (±0.028)	86.45% (±0.056)	85.91% (±0.032)	25.32% (±0.033)	7.71% (±0.034)
GAT-ClustGCN	✓		93.29% (±0.075)	94.05% (±0.043)	92.18% (±0.071)	19.21% (±0.062)	6.02% (±0.039)	92.81% (±0.027)	93.17% (±0.041)	**89.61%** (±0.026)	22.13% (±0.033)	7.16% (±0.049)
		✓	93.16% (±0.037)	93.75% (±0.072)	**92.42%** (±0.057)	19.09% (±0.081)	6.41% (±0.047)	92.59% (±0.041)	93.08% (±0.032)	89.24% (±0.039)	22.31% (±0.043)	7.63% (±0.055)

10.5. Transductive vs. Inductive Learning

In this final test case, we investigated the efficacy of the transductive against the inductive learning schemes (Section 8.1). Table 2 presents detailed results pertaining to both *DMA-GCN(CT/SEQ)* structures.

Table 2. Summary of segmentation performance measures (means ± stds) of the two cartilage classes of our proposed methods *DMA-GCN (CT/SEQ)* by varying the overall learning paradigm (transductive vs. inductive). Best results for each category (*CT* vs *SEQ*) with respect to *DSC* index are highlighted.

Module	SEQ	CT	Femoral Cartilage					Tibial Cartilage				
			Recall	Precision	$\mathcal{DSC}$	$\mathcal{VOE}$	$\mathcal{VD}$	Recall	Precision	$\mathcal{DSC}$	$\mathcal{VOE}$	$\mathcal{VD}$
Inductive	✓		91.78% (±0.051)	89.61% (±0.038)	**89.45%** (±0.086)	15.11% (±0.093)	5.98% (±0.067)	86.09% (±0.036)	84.29% (±0.053)	83.81% (±0.041)	26.02% (±0.094)	8.45% (±0.078)
		✓	85.78% (±0.045)	87.71% (±0.033)	85.87% (±0.056)	23.61% (±0.069)	7.79% (±0.041)	85.92% (±0.049)	84.93% (±0.071)	**84.05%** (±0.046)	26.15% (±0.029)	8.71% (±0.027)
Transductive	✓		96.13% (±0.121)	95.21% (±0.183)	95.44% (±0.022)	13.21% (±0.045)	3.98% (±0.089)	95.19% (±0.031)	94.41% (±0.027)	93.87% (±0.024)	18.65% (±0.034)	5.03% (±0.023)
		✓	96.17% (±0.117)	95.81% (±0.189)	**95.71%** (±0.032)	13.17% (±0.051)	3.94% (±0.094)	95.31% (±0.032)	94.78% (±0.025)	**94.02%** (±0.021)	17.98% (±0.034)	4.99% (±0.029)

According to the results, transductive learning significantly outperforms the inductive learning scenario, for both cartilage classes of interest and across all evaluations metrics. This can be attributed to the following reasons. First, corroborating the well-established finding of the literature, the superior rates underscore the importance of utilizing the *SSL* features of the unlabeled nodes in the training process, combined with those of labeled ones. Secondly, the refreshing learning stage applied in mini-batch learning (Section 10.3) allows the network to appropriately adjust to the newly observed out-of-sampling batches. On the other hand, the inductive model is trained once using the nearest neighbor image. This network is then used to segment the target image, by classifying the entire set of unlabeled batches from $\mathcal{T}$.

10.6. Comparative Results

Table 3 presents extensive comparative results, contrasting our *DMA-GCN* models with traditional patch-based approaches, and state-of-the-art deep learning architectures established in the field of medical image segmentation. We also applied six graph-based convolution networks in the comparisons. These networks were used as standalone models solely to conduct global learning of the nodes. Finally, we applied the more integrated *MGCCN* model [30]. For the *DMA-GCN* models, we applied the following parameter setting: $N_{\mathcal{A}} = 10$ atlases, $K = 8$ attention heads, $S = 2$ spatial scales, $M = 3$ dense layers, and transductive learning with a mini-batch size of 128.

Based on the results of Table 3, we should notice the more enhanced rates of *GAT* and *MGCN* compared with those of the other graph convolution networks, suggesting that the attention mechanism combined with a multi-scale consideration of the data can improve the model's performance. Furthermore, the proposed deep *DMA-GCN(CT)* and *DMA-GCN(SEQ)* models are both shown to outperform all competing methods in the experimental setup, achieving $\mathcal{DSC}_{fmrl} = (95.71\%, 95.44\%)$ and $\mathcal{DSC}_{tbl} = (94.02\%, 93.87\%)$, respectively, across all evaluation metrics and in both femoral and tibial segmentation. *DMA-GCN(CT)* provides a slightly better performance compared to the *DMA-GCN(SEQ)* indicating that the alternating combination of local–global convolutional units is more effective. However, both methods may fail to deliver satisfactory results in certain cases where the cartilage tissue is severely damaged or otherwise deformed. Figure 18 showcases an example of a successful application of both *DMA-GCN(SEQ)* and *DMA-GCN(CT)* models, along with a marginal case exhibiting suboptimal results, due to extreme cartilage thinning.

Table 3. Summary of segmentation performance measures (means ± stds) of the two cartilage classes of our proposed methods *DMA-GCN (CT)* and *DMA-GCN (SEQ)* compared to state of the art: 1. patch-based methods, 2. deep learning methods, 3. graph deep learning methods. Best results for all three categories (*CT* vs *SEQ*) with respect to *DSC* index are highlighted.

	Femoral Cartilage					**Tibial Cartilage**				
Method	***Recall***	***Precision***	$\mathcal{DSC}$	$\mathcal{VOE}$	$\mathcal{VD}$	**Recall**	**Precision**	$\mathcal{DSC}$	$\mathcal{VOE}$	$\mathcal{VD}$
PB_{SC}	83.51% (±0.066)	82.65% (±0.045)	82.23% (±0.061)	29.97% (±0.082)	11.01% (±0.037)	79.76% (±0.021)	81.45% (±0.036)	78.85% (±0.027)	34.28% (±0.031)	11.79% (±0.041)
PB_{NLM}	84.12% (±0.071)	83.28% (±0.045)	84.09% (±0.052)	26.73% (±0.061)	8.25% (±0.084)	81.22% (±0.038)	82.07% (±0.019)	80.04% (±0.027)	33.91% (±0.031)	11.41% (±0.029)
HyLP	94.04% (±0.052)	93.16% (±0.051)	**92.56%** (±0.026)	15.16% (±0.028)	5.12% (±0.034)	91.08% (±0.074)	89.98% (±0.023)	**89.91%** (±0.018)	19.67% (±0.025)	5.85% (±0.011)
SegNet	89.18% (±0.116)	89.48% (±0.219)	**89.09%** (±0.089)	20.73% (±0.039)	5.65% (±0.066)	87.22% (±0.056)	89.07% (±0.062)	**86.12%** (±0.034)	22.79% (±0.012)	6.23% (±0.016)
DenseVoxNet	88.75% (±0.156)	88.67% (±0.204)	87.54% (±0.042)	21.83% (±0.048)	6.45% (±0.121)	87.45% (±0.076)	86.03% (±0.041)	85.68% (±0.047)	25.47% (±0.028)	7.98% (±0.025)
VoxResNet	88.03% (±0.187)	88.92% (±0.205)	88.12% (±0.047)	22.71% (±0.055)	6.64% (±0.128)	87.04% (±0.071)	85.26% (±0.044)	85.12% (±0.043)	26.03% (±0.032)	8.04% (±0.029)
KCB-Net	89.74% (±0.149)	90.12% (±0.185)	88.92% (±0.031)	23.13% (±0.042)	6.72% (±0.098)	88.12% (±0.055)	87.46% (±0.029)	87.92% (±0.033)	25.90% (±0.017)	8.04% (±0.021)
CAN3D	88.04% (±0.156)	88.54% (±0.205)	87.12% (±0.03)	22.93% (±0.042)	6.59% (±0.098)	87.28% (±0.055)	85.26% (±0.029)	85.02% (±0.033)	25.76% (±0.017)	8.01% (±0.021)
PointNet	87.13% (±0.121)	88.12% (±0.187)	86.49% (±0.023)	23.06% (±0.051)	7.13% (±0.104)	86.88% (±0.062)	85.03% (±0.031)	84.79% (±0.023)	24.92% (±0.024)	7.39% (±0.018)
GCN	90.19% (±0.129)	90.84% (±0.126)	89.23% (±0.030)	19.65% (±0.046)	5.48% (±0.098)	88.92% (±0.059)	89.02% (±0.033)	88.26% (±0.021)	23.27% (±0.028)	6.78% (±0.017)
SGC	91.02% (±0.212)	91.31% (±0.132)	89.84% (±0.032)	17.41% (±0.064)	5.19% (±0.098)	89.81% (±0.061)	89.54% (±0.047)	89.02% (±0.032)	22.11% (±0.039)	5.89% (±0.028)
ClusterGCN	90.56% (±0.141)	91.08% (±0.150)	90.12% (±0.039)	17.33% (±0.058)	5.16% (±0.107)	90.28% (±0.054)	91.05% (±0.061)	89.93% (±0.039)	22.08% (±0.036)	5.82% (±0.024)
GraphSAINT	92.61% (±0.132)	92.74% (±0.131)	90.87% (±0.027)	17.18% (±0.051)	5.16% (±0.102)	91.75% (±0.054)	91.04% (±0.029)	90.12% (±0.026)	22.04% (±0.022)	5.76% (±0.020)
GraphSAGE	92.87% (±0.129)	92.91% (±0.144)	90.95% (±0.031)	17.12% (±0.065)	5.09% (±0.098)	92.04% (±0.049)	92.53% (±0.033)	90.49% (±0.038)	20.71% (±0.027)	5.66% (±0.024)
GAT	93.14% (±0.141)	93.09% (±0.203)	92.87% (±0.045)	13.90% (±0.051)	4.36% (±0.101)	93.29% (±0.055)	93.81% (±0.031)	90.86% (±0.019)	19.58% (±0.029)	5.61% (±0.026)
MGCN	94.11% (±0.125)	93.92% (±0.181)	**93.27%** (±0.029)	14.05% (±0.045)	4.27% (±0.096)	93.79% (±0.041)	94.06% (±0.027)	**91.43%** (±0.024)	19.84% (±0.035)	5.34% (±0.021)
DMA-GCN (SEQ)	96.13% (±0.121)	95.21% (±0.183)	95.44% (±0.022)	13.21% (±0.045)	3.98% (±0.089)	95.19% (±0.031)	94.41% (±0.027)	93.87% (±0.024)	18.65% (±0.034)	5.03% (±0.023)
DMA-GCN (CT)	96.17% (±0.117)	95.81% (±0.189)	**95.71%** (±0.032)	13.17% (±0.051)	3.94% (±0.094)	95.31% (±0.032)	94.78% (±0.025)	**94.02%** (±0.021)	17.98% (±0.034)	4.99% (±0.029)

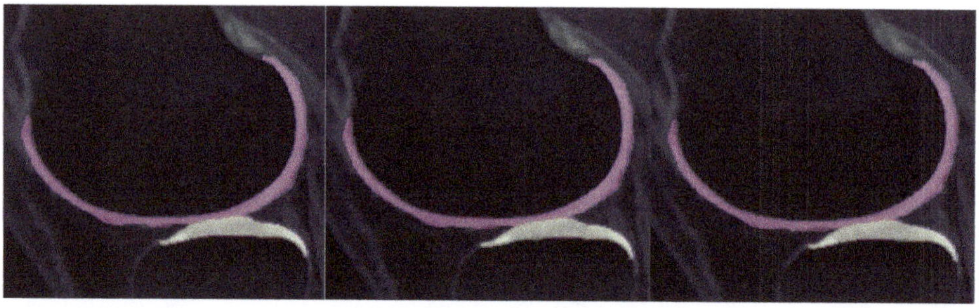

(**a**)

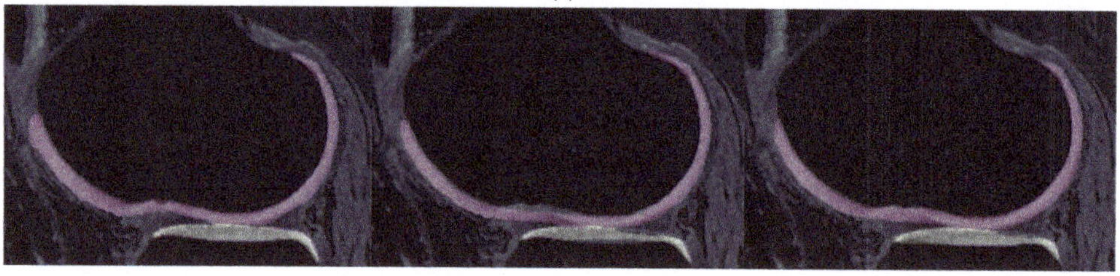

(**b**)

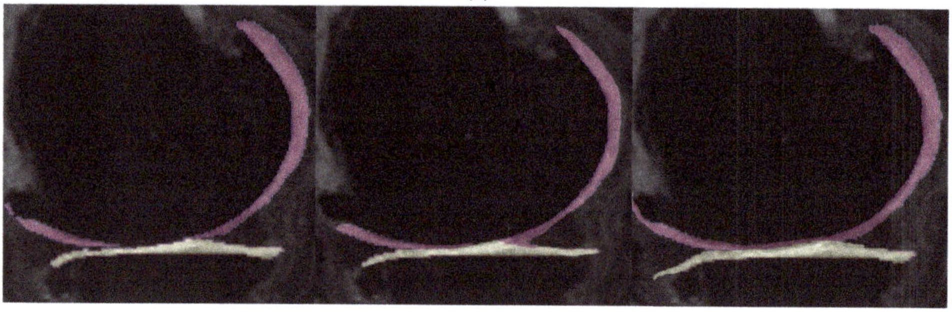

(**c**)

Figure 18. Segmentation results for femoral (FC) and tibial (TC) cartilage for the two main proposed models (*DMA-GCN(SEQ)* and *DMA-GCN(CT)*). The first part of the figure illustrates a case of successful application of *DMA-GCN* on a healthy knee (KL grade 0), while the second and third parts correspond to more challenging subjects with moderate (KL grade 2) and severe (KL grade 4) osteoarthritis. (Left to right: ground truth, *DMA-GCN(SEQ)*, *DMA-CGN(CT)*—color coding: pink → FC, white → TC). (**a**) Segmentation showcase—KL grade 0. (**b**) Segmentation showcase—KL grade 2. (**c**) Segmentation showcase—KL grade 4.

11. Conclusions and Future Work

In this paper, we presented the *DMA-GCN* for knee joint cartilage segmentation. Our models shared a number of attractive properties, such as the constructive integration of local-level and global-level learning, a densely connected structure, and adaptive graph learning. These features rendered the *DMA-GCN* capable of acquiring expensive node representations. A comparative analysis with various state-of-the-art deep learning and graph-based convolution networks validated the efficacy of the proposed approach. As future research, we intend to extent the current framework by focusing on hypergraph networks, which allow the incorporation of multiple views of graph data.

Author Contributions: Conceptualization, J.T.; methodology, C.C. and J.T.; software, C.C.; writing—original draft preparation, C.C.; writing—review and editing, J.T., D.T, A.S., S.M. and C.C.; visualization, C.C. and S.M.; supervision, J.T. All authors have read and agreed to the published version of the manuscript.

Funding: This research received no external funding.

Data Availability Statement: The data presented in this study are available in [4,49].

Conflicts of Interest: The authors declare that they have no conflicts of interest or personal relationships that could have appeared to influence the work reported in this paper.

References

1. Felipe, J.; McCombie, J. Burden of major musculoskeletal conditions. *ERD Work. Pap. Ser.* **2002**, *81*, 1–27.
2. Ebrahimkhani, S.; Jaward, M.H.; Cicuttini, F.M.; Dharmaratne, A.; Wang, Y.; de Herrera, A.G. A review on segmentation of knee articular cartilage: From conventional methods towards deep learning. *Artif. Intell. Med.* **2020**, *106*, 101851. [CrossRef] [PubMed]
3. Fripp, J.; Crozier, S.; Warfield, S.K.; Ourselin, S. Automatic segmentation of the bone and extraction of the bone-cartilage interface from magnetic resonance images of the knee. *Phys. Med. Biol.* **2007**, *52*, 1617–1631. [CrossRef]
4. Ambellan, F.; Tack, A.; Ehlke, M.; Zachow, S. Automated segmentation of knee bone and cartilage combining statistical shape knowledge and convolutional neural networks: Data from the Osteoarthritis Initiative. *Med. Image Anal.* **2019**, *52*, 109–118. [CrossRef] [PubMed]
5. Folkesson, J.; Dam, E.B.; Olsen, O.F.; Pettersen, P.C.; Christiansen, C. Segmenting articular cartilage automatically using a voxel classification approach. *IEEE Trans. Med. Imaging* **2007**, *26*, 106–115. [CrossRef] [PubMed]
6. Zhang, K.; Lu, W.; Marziliano, P. Automatic knee cartilage segmentation from multi-contrast MR images using support vector machine classification with spatial dependencies. *Magn. Reson. Imaging* **2013**, *31*, 1731–1743. [CrossRef] [PubMed]
7. Rousseau, F.; Habas, P.A.; Studholme, C. A supervised patch-based approach for human brain labeling. *IEEE Trans. Med. Imaging* **2011**, *30*, 1852–1862. [CrossRef]
8. Zhang, D.; Guo, Q.; Wu, G.; Shen, D. Sparse patch-based label fusion for multi-atlas segmentation. In *Multimodal Brain Image Analysis, Proceedings of the Multimodal Brain Image Analysis: Second International Workshop, MBIA 2012, Nice, France, 1–5 October 2012*; Lecture Notes in Computer Science Series ; Springer: Berlin/Heidelberg, Germany, 2012; Volume 7509, pp. 94–102. [CrossRef]
9. Hajnal, J.V.; Hill, D.L.; Hawkes, D.J. Medical image registration. *Med. Image Regist.* **2001**, *46*, 1–383. [CrossRef]
10. Wang, R.; Lei, T.; Cui, R.; Zhang, B.; Meng, H.; Nandi, A.K. Medical image segmentation using deep learning: A survey. *IET Image Process.* **2022**, *16*, 1243–1267. [CrossRef]
11. Minaee, S.; Boykov, Y.; Porikli, F.; Plaza, A.; Kehtarnavaz, N.; Terzopoulos, D. Image Segmentation Using Deep Learning: A Survey. *IEEE Trans. Pattern Anal. Mach. Intell.* **2022**, *44*, 3523–3542. [CrossRef]
12. Badrinarayanan, V.; Kendall, A.; Cipolla, R. SegNet: A Deep Convolutional Encoder-Decoder Architecture for Image Segmentation. *IEEE Trans. Pattern Anal. Mach. Intell.* **2017**, *39*, 2481–2495. [CrossRef] [PubMed]
13. Yu, L.; Cheng, J.Z.; Dou, Q.; Yang, X.; Chen, H.; Qin, J.; Heng, P.A. Automatic 3D cardiovascular MR segmentation with densely-connected volumetric convnets. In *Medical Image Computing and Computer-Assisted Intervention—MICCAI 2017, Proceedings of the 20th International Conference, Quebec City, QC, Canada, 11–13 September 2017*; Lecture Notes in Computer Science Series ; Springer: Cham, Switzerland, 2017; Volume 10434, pp. 287–295. [CrossRef]
14. Chen, H.; Dou, Q.; Yu, L.; Heng, P.A. VoxResNet: Deep Voxelwise Residual Networks for Volumetric Brain Segmentation. *arXiv* **2016**, arXiv:1608.05895.
15. Qi, C.R.; Su, H.; Mo, K.; Guibas, L.J. PointNet: Deep learning on point sets for 3D classification and segmentation. In Proceedings of the 30th IEEE Conference on Computer Vision and Pattern Recognition, CVPR 2017, Honolulu, HI, USA, 21–26 July 2017; pp. 77–85. [CrossRef]
16. Xie, H.; Pan, Z.; Zhou, L.; Zaman, F.A.; Chen, D.Z.; Jonas, J.B.; Xu, W.; Wang, Y.X.; Wu, X. Globally optimal OCT surface segmentation using a constrained IPM optimization. *Opt. Express* **2022**, *30*, 2453. [CrossRef] [PubMed]
17. Wu, Z.; Pan, S.; Chen, F.; Long, G.; Zhang, C.; Yu, P.S. A Comprehensive Survey on Graph Neural Networks. *IEEE Trans. Neural Netw. Learn. Syst.* **2021**, *32*, 4–24. [CrossRef] [PubMed]
18. Defferrard, M.; Bresson, X.; Vandergheynst, P. Convolutional Neural Networks on Graphs with Fast Localized Spectral Filtering. *Adv. Neural Inf. Process. Syst.* **2016**, *29*, 395–398.
19. Kipf, T.N.; Welling, M. Semi-supervised classification with graph convolutional networks. In Proceedings of the 5th International Conference on Learning Representations, ICLR 2017—Conference Track Proceedings, Toulon, France, 24–26 April 2017; pp. 1–14.
20. Gilmer, J.; Schoenholz, S.S.; Riley, P.F.; Vinyals, O.; Dahl, G.E. Neural Message Passing for Quantum Chemistry. In Proceedings of the 34th International Conference on Machine Learning, Sydney, Australia, 6–11 August 2017.
21. Zeng, H.; Zhou, H.; Srivastava, A.; Kannan, R.; Prasanna, V. GraphSAINT: Graph sampling based inductive learning method. In Proceedings of the 8th International Conference on Learning Representations, ICLR 2020, Addis Ababa, Ethiopia, 26–30 April 2020; pp. 1–19.
22. Chen, J.; Ma, T.; Xiao, C. FastGCN: Fast learning with graph convolu-tional networks via importance sampling. In Proceedings of the 6th International Conference on Learning Representations, ICLR 2018—Conference Track Proceedings, Vancouver, BC, Canada, 30 April–3 May 2018; pp. 1–15.

23. Chiang, W.L.; Li, Y.; Liu, X.; Bengio, S.; Si, S.; Hsieh, C.J. Cluster-GCN: An efficient algorithm for training deep and large graph convolutional networks. In Proceedings of the ACM SIGKDD International Conference on Knowledge Discovery and Data Mining, Anchorage AK USA, 4–8 August 2019; pp. 257–266. [CrossRef]
24. Veličković, P.; Casanova, A.; Liò, P.; Cucurull, G.; Romero, A.; Bengio, Y. Graph attention networks. In Proceedings of the 6th International Conference on Learning Representations, ICLR 2018—Conference Track Proceedings, Vancouver, BC, Canada, 30 April–3 May 2018; pp. 1–12. [CrossRef]
25. Hamilton, W.L.; Ying, R.; Leskovec, J. Inductive representation learning on large graphs. *Adv. Neural Inf. Process. Syst.* **2017**, *30*, 1025–1035.
26. Qiu, J.; Tang, J.; Ma, H.; Dong, Y.; Wang, K.; Tang, J. DeepInf: Social influence prediction with deep learning. In Proceedings of the ACM SIGKDD International Conference on Knowledge Discovery and Data Mining, London, UK, 19–23 August 2018; pp. 2110–2119. [CrossRef]
27. Science, C. A Graph-to-Sequence Model for AMR-to-Text Generation. *arXiv* **2017**, arXiv:1805.02473v3.
28. Cao, N.D.; Kipf, T. MolGAN: An implicit generative model for small molecular graphs. *arXiv* **2019**, arXiv:1805.11973v2.
29. Yan, S.; Xiong, Y.; Lin, D. Spatial Temporal Graph Convolutional Networks for Skeleton-Based Action Recognition. *arXiv* **2017**, arXiv:1801.07455v2.
30. Wan, S.; Pan, S.; Zhong, S.; Yang, J.; Yang, J.; Zhan, Y.; Gong, C. Multi-level graph learning network for hyperspectral image classification. *Pattern Recognit.* **2022**, *129*, 108705. [CrossRef]
31. Yang, P.; Tong, L.; Qian, B.; Gao, Z.; Yu, J.; Xiao, C. Hyperspectral Image Classification with Spectral and Spatial Graph Using Inductive Representation Learning Network. *IEEE J. Sel. Top. Appl. Earth Obs. Remote Sens.* **2021**, *14*, 791–800. [CrossRef]
32. Jia, S.; Jiang, S.; Zhang, S.; Xu, M.; Jia, X. Graph-in-Graph Convolutional Network for Hyperspectral Image Classification. *IEEE Trans. Neural Netw. Learn. Syst.* **2022**, *35*, 1157–1171. [CrossRef] [PubMed]
33. Zhang, M.; Chen, Y. Link prediction based on graph neural networks. *Adv. Neural Inf. Process. Syst.* **2018**, *31*, 5165–5175.
34. Yu, B.; Yin, H.; Zhu, Z. Spatio-temporal graph convolutional networks: A deep learning framework for traffic forecasting. *IJCAI Int. Jt. Conf. Artif. Intell.* **2018**, *2018*, 3634–3640.
35. Guo, Z.; Li, X.; Huang, H.; Guo, N.; Li, Q. Deep learning-based image segmentation on multimodal medical imaging. *IEEE Trans. Radiat. Plasma Med. Sci.* **2019**, *3*, 162–169. [CrossRef] [PubMed]
36. Ma, Y.; Tang, J. *Deep Learning on Graphs*; Cambridge University Press: Cambridge, UK, 2021.
37. Chadoulos, C.; Moustakidis, S.; Tsaopoulos, D.; Theocharis, J. Multi-atlas segmentation of knee cartilage by Propagating Labels via Semi-supervised Learning. In Proceedings of the IMIP 2022: 2022 4th International Conference on Intelligent Medicine and Image Processing, Tianjin, China, 18–21 March 2022; pp. 76–82. [CrossRef]
38. Peterfy, C.G.; Schneider, E.; Nevitt, M. The osteoarthritis initiative: Report on the design rationale for the magnetic resonance imaging protocol for the knee. *Osteoarthr. Cartil.* **2008**, *16*, 1433–1441. [CrossRef]
39. Sethian, J. Advancing interfaces: Level set and fast marching methods. In *Level Set Methods and Fast Marching Methods*, 2nd ed.; Cambridge Press: Cambridge, UK, 1999; Chapter 16, p. 12.
40. Sled, J.G.; Zijdenbos, A.P.; Evans, A.C. A nonparametric method for automatic correction of intensity nonuniformity in mri data. *IEEE Trans. Med. Imaging* **1998**, *17*, 87–97. [CrossRef]
41. Nyul, L.G.; Udupa, J.K. Standardizing the MR image intensity scales: Making MR intensities have tissue specific meaning. *Med. Imaging 2000 Image Disp. Vis.* **2000**, *3976*, 496–504.
42. Buades, A.; Coll, B.; Morel, J.M. Non-Local Means Denoising. *Image Process. Line* **2011**, *1*, 208–212. [CrossRef]
43. Dalal, N.; Triggs, B. Histogram of Oriented Gradients for Human Detection. *IEEE Trans. Ind. Informatics* **2020**, *16*, 4714–4725. [CrossRef]
44. Klaser, A.; Marszalek, M.; Schmid, C. A spatio-temporal descriptor based on 3D-gradients. In Proceedings of the BMVC 2008—Proceedings of the British Machine Vision Conference 2008, Leeds, UK, Sepember 2008.
45. Liu, Q.; Xiao, L.; Yang, J.; Wei, Z. CNN-Enhanced Graph Convolutional Network with Pixel- and Superpixel-Level Feature Fusion for Hyperspectral Image Classification. *IEEE Trans. Geosci. Remote Sens.* **2021**, *59*, 8657–8671. [CrossRef]
46. Simonyan, K.; Zisserman, A. Very deep convolutional networks for large-scale image recognition. In Proceedings of the 3rd International Conference on Learning Representations, ICLR 2015—Conference Track Proceedings, San Diego, CA, USA, 7–9 May 2015; pp. 1–14.
47. Peng, Y.; Zheng, H.; Liang, P.; Zhang, L.; Zaman, F.; Wu, X.; Sonka, M.; Chen, D.Z. KCB-Net: A 3D knee cartilage and bone segmentation network via sparse annotation. *Med. Image Anal.* **2022**, *82*, 102574. [CrossRef] [PubMed]
48. Jaderberg, M.; Simonyan, K.; Zisserman, A.; Kavukcuoglu, K. Spatial transformer networks. *Adv. Neural Inf. Process. Syst.* **2015**, *28*, 2017–2025.
49. Schneider, E.; NessAiver, M.; White, D.; Purdy, D.; Martin, L.; Fanella, L.; Davis, D.; Vignone, M.; Wu, G.; Gullapalli, R. The osteoarthritis initiative (OAI) magnetic resonance imaging quality assurance methods and results. *Osteoarthr. Cartil.* **2008**, *16*, 994–1004. [CrossRef]

bioengineering

MDPI

Article

Hybrid Deep Neural Network Framework Combining Skeleton and Gait Features for Pathological Gait Recognition

Kooksung Jun [1,2], Keunhan Lee [3], Sanghyub Lee [2], Hwanho Lee [3,*] and Mun Sang Kim [2,*]

1 Robocare, Seongnam 13449, Republic of Korea; ks_jun@robocare.co.kr
2 School of Integrated Technology, Gwangju Institute of Science and Technology, Gwangju 61005, Republic of Korea; sang-hyub@gist.ac.kr
3 Department of Otolaryngology-Head and Neck Surgery, Kosin University College of Medicine, Busan 49267, Republic of Korea; aya051@naver.com
* Correspondence: hornet999@hanmail.net (H.L.); munsang@gist.ac.kr (M.S.K.)

Abstract: Human skeleton data obtained using a depth camera have been used for pathological gait recognition to support doctor or physician diagnosis decisions. Most studies for skeleton-based pathological gait recognition have used either raw skeleton sequences directly or gait features, such as gait parameters and joint angles, extracted from raw skeleton sequences. We hypothesize that using skeleton, joint angles, and gait parameters together can improve recognition performance. This study aims to develop a deep neural network model that effectively combines different types of input data. We propose a hybrid deep neural network framework composed of a graph convolutional network, recurrent neural network, and artificial neural network to effectively encode skeleton sequences, joint angle sequences, and gait parameters, respectively. The features extracted from three different input data types are fused and fed into the final classification layer. We evaluate the proposed model on two different skeleton datasets (a simulated pathological gait dataset and a vestibular disorder gait dataset) that were collected using an Azure Kinect. The proposed model, with multiple types of input, improved the pathological gait recognition performance compared to single input models on both datasets. Furthermore, it achieved the best performance among the state-of-the-art models for skeleton-based action recognition.

Keywords: hybrid deep neural network; feature fusion; pathological gait recognition; skeleton-based gait analysis

Citation: Jun, K.; Lee, K.; Lee, S.; Lee, H.; Kim, M.S. Hybrid Deep Neural Network Framework Combining Skeleton and Gait Features for Pathological Gait Recognition. *Bioengineering* **2023**, *10*, 1133. https://doi.org/10.3390/bioengineering10101133

Academic Editors: Hongqing Yu, Alaa AlZoubi, Yifan Zhao and Hongbo Du

Received: 8 August 2023
Revised: 15 September 2023
Accepted: 19 September 2023
Published: 27 September 2023

1. Introduction

Gait represents crucial bioinformation, necessitating the proper integration of sensory, motor, and cognitive functions, and has consequently been the subject of extensive investigation for a considerable duration [1,2]. If the weakness of some body parts has a negative influence on those functions, the gait pattern can become abnormal and unbalanced. In other words, abnormal and unbalanced gait patterns indicate disorders of some body functions, and it is possible to find them by analyzing gait. Parkinson's disease is a prominent example of a condition that manifests abnormal gait patterns, with numerous studies conducted on its gait characteristics [3–6]. It is marked by symptoms including a gradual reduction in walking speed, diminished swinging motion of the arms, shorter stride length, impaired balance, and a decline in the coordination of arm and trunk movements during walking. Additionally, numerous studies have explored the relationship between gait patterns and other specific diseases, including autism spectrum disorder [7,8], stroke [9,10], Alzheimer's disease [11], vestibular problems [12–14], and functional gait disorders [15,16]. Furthermore, gait patterns have been used in practice to support doctor or physician decisions for patients. There are many research groups studying gait patterns and various research results continue to be published.

Recognizing pathological gait patterns helps to diagnose a disease and even to find a presymptom of a disease before it worsens. Therefore, there have been many approaches for pathological gait recognition using various sensors, such as inertial measurement units (IMUs), plantar foot pressure sensors, motion capture systems, and depth cameras. Sensor-based systems have many advantages. They make it possible to automatically prescreen for specific diseases without visiting a hospital. People hardly realize whether their gait patterns are changed or not because they change gradually. It might be too late if they realize their abnormal gait by themselves. On the other hand, sensor systems can analyze a gait pattern with objective standards, so it is possible to detect abnormal gaits in the early phase of a disease, and patients can receive proper treatment before the disease worsens. Therefore, if a sensor-based pathological gait recognition system is installed in a home or elderly care center and people conduct gait analysis periodically, specific diseases can be prescreened without visiting a hospital.

A depth camera was used to recognize pathological gaits in this study. A depth camera, such as Kinect (Microsoft Corp., Redmond, WA, USA), Astra (Orbbec 3D Technology International, Inc., Troy, MI, USA), and Realsense (Intel Corp., Santa Clara, CA, USA), can obtain not only RGB data but also depth data for each pixel. The collected RGB and depth data can be used to simulate the human skeleton, which contains three-dimensional positional information of each joint. A depth camera can measure gait data without attaching sensors or markers, whereas an IMU and motion capture system require the attachment of sensors or markers, which can make the walker feel uncomfortable and walk unnaturally. Furthermore, a depth camera can obtain information on all body joints. A depth camera-based gait analysis system is simple to operate and has a relatively low cost and reasonable accuracy, so it can be operated in various environments. Therefore, many studies have used human skeleton data obtained through a depth camera to recognize pathological gaits [3,4,11,17–27].

In the domain of skeleton-based pathological recognition, numerous research studies have been conducted utilizing machine learning algorithms. For instance, Li et al. [3] proposed a method to classify normal individuals and patients with hemiplegia and Parkinson's disease using k-nearest neighbors. They used a covariance matrix representing joint motions and speeds extracted from the skeleton sequence. Dranca et al. [4] introduced a machine learning-based method to classify Parkinson's disease stages. They extracted features from the skeleton by applying correlation-based feature selection, information gain, or consistency subset evaluation. Seifallahi et al. [11] proposed a method to detect Alzheimer's disease by employing a support vector machine (SVM) classifier with a Gaussian kernel. Gait parameters, such as time walking, step length, step number, stride length, gait cycle, and stride velocity, were fed to the classification model in their work. Bei et al. [22] proposed a method to detect movement disorders using machine learning algorithms. Gait parameters, such as gait symmetry, step length, and gait cycle, were fed to the classification model in their study. Chakraborty et al. [23] introduced a method for the automatic diagnosis of cerebral palsy gait using a multi-Kinect system and SVM-based classification model. They extracted spatiotemporal features from the skeleton and used them as the input data to the classifier. Chakraborty et al. [24] employed a multiple adaptive regression splines model to recognize equinus foot deformity gait. The hip, knee, and ankle angles of both sides were extracted from the skeleton and used as the input data in their work.

Deep neural network models have also been applied to skeleton-based pathological gait recognition. For instance, Guo et al. [17] proposed a bidirectional long short-term memory (LSTM)-based model to classify normal, in-toeing, out-toeing, drop-foot, pronation, and supination gaits. The lower limb skeleton was used to extract statistical features and angle sequences in their study. Tian et al. [25] proposed a spatiotemporal attention-enhanced gait-structural graph convolutional network (AGS-GCN) to recognize abnormal gaits. They used the lower limb joints and spine base to extract spatiotemporal gait parameters, such as the joint trajectory, joint angle, and gait link. Sadeghzadehyazdi et al. [26] proposed a hybrid model composed of a convolutional neural network (CNN) and LSTM to model spatiotemporal patterns for gait anomaly recognition. They used normalized joints for the classification. Kim et al. [27] applied a spatiotemporal GCN with an attention mechanism to the spatiotemporal features extracted from skeleton data for the recognition of abnormal gaits.

Most existing methods for skeleton-based pathological gait recognition use gait features extracted from raw skeletons, such as static gait parameters [11,22,23] and joint angle sequences [4,17,24]. They have shown their effectiveness in recognizing pathological gaits. Gait features effectively represent gait abnormalities, so they can be interpreted more easily than raw skeleton data. However, this does not mean that they can represent all the important information of the raw skeleton data. On the other hand, the raw skeleton data include all the important information; however, it is difficult to understand a gait abnormality because of the complicated structure and large data size. Recently, studies inputting the skeleton itself into a model have been published [28–30]. They showed the possibility of recognizing pathological gaits by interpreting the skeleton data, but gait features were not considered at all. Prior studies have not extensively explored a hybrid approach that combines the advantages of both gait features and raw skeleton data. This means that while gait features provide valuable insights into gait abnormalities, they may not fully exploit the richness of information contained in the raw skeleton data. Thus, there is an opportunity to investigate novel methods that leverage both gait features and raw skeleton data to enhance the recognition of pathological gaits. Following this observation, our motivation evolved into exploring novel methods that harness the complementary strengths of both gait features and raw skeleton data to improve pathological gait recognition.

We hypothesized that using both gait features and raw skeleton data could further improve the performance of pathological gait recognition since they have different advantages and representations. Gait parameters are the most compressed form to effectively represent the abnormality of a gait, joint sequences are focused on showing the bending and balancing abilities of lower limb joints, and raw skeleton sequences preserve all important information for pathological gait recognition and show the overall movement of the whole body during walking. Using these together can facilitate a model to converge to the global minima since gait abnormalities can be interpreted through a variety of perspectives. However, a method to use all of them together for pathological gait recognition has not yet been proposed. In consideration of these factors, we have innovatively introduced a novel methodology encompassing the concurrent utilization of gait parameters, joint sequences, and raw skeleton sequences, marking a pioneering advancement in the field.

In this paper, we propose a novel hybrid deep learning model designed to maximize the utilization of raw skeleton data and gait features for pathological gait recognition. Since the input data have different characteristics, we applied different deep learning architectures to encode each input data effectively. A graph convolutional network (GCN), recurrent neural network (RNN), and artificial neural network (ANN) are used to encode the raw skeleton sequences, the joint angle sequences, and the gait parameters, respectively. Their outputs are fused together and fed into the final classification layer. This fusion of features can be achieved through concatenation, with further performance enhancements achievable through feature selection or weighting techniques. This involves the selection or assignment of weights to features and matching scores that demonstrate low correlation, as exemplified by [31], and high discrimination, as illustrated by [32].

The primary objective of this study is to demonstrate improved performance in recognizing pathological gait patterns through the fusion of raw skeleton data and gait features with our hybrid deep learning model designed to synergize diverse input types. Given the inherent diversity in the characteristics of input data, our approach incorporates distinct deep learning architectures tailored for encoding each specific data type. Our proposed model stands as an innovative contribution by integrating raw skeleton sequences, joint angle sequences, and gait parameters, marking the first of its kind in the realm of pathological gait recognition. To substantiate the efficacy of this pioneering model, comprehensive evaluations were conducted on diverse pathological gait datasets, including a simulated pathological gait dataset and a vestibular disorder gait dataset, both meticulously collected utilizing Azure Kinect. Furthermore, rigorous comparative analyses were conducted against state-of-the-art models specialized in skeleton-based action recognition.

2. Materials and Methods

Most studies in the field of skeleton-based pathological gait recognition have traditionally focused on two primary approaches: utilizing raw skeleton sequences directly or extracting gait features, such as gait parameters and joint angles, from these raw sequences. However, these studies have often treated each data type in isolation or separately. In this study, we aim to address the potential for improved recognition performance by effectively combining different types of input data. To achieve this, we propose a novel deep neural network model. This model adopts a hybrid deep neural network framework, consisting of GCN, RNN, and ANN layers. Each of these components is specifically designed to encode skeleton sequences, joint angle sequences, and gait parameters, respectively. The extracted features from these three distinct data types are then fused together and input into the final classification layer. A comprehensive illustration of this network is presented in Figure 1. By employing this innovative approach, our research aims to enhance pathological gait recognition performance. To demonstrate the effectiveness of our proposed model, we collected two skeleton datasets using Azure Kinect: a simulated pathological gait dataset and a vestibular disorder gait dataset, and subsequently conducted evaluations using these datasets.

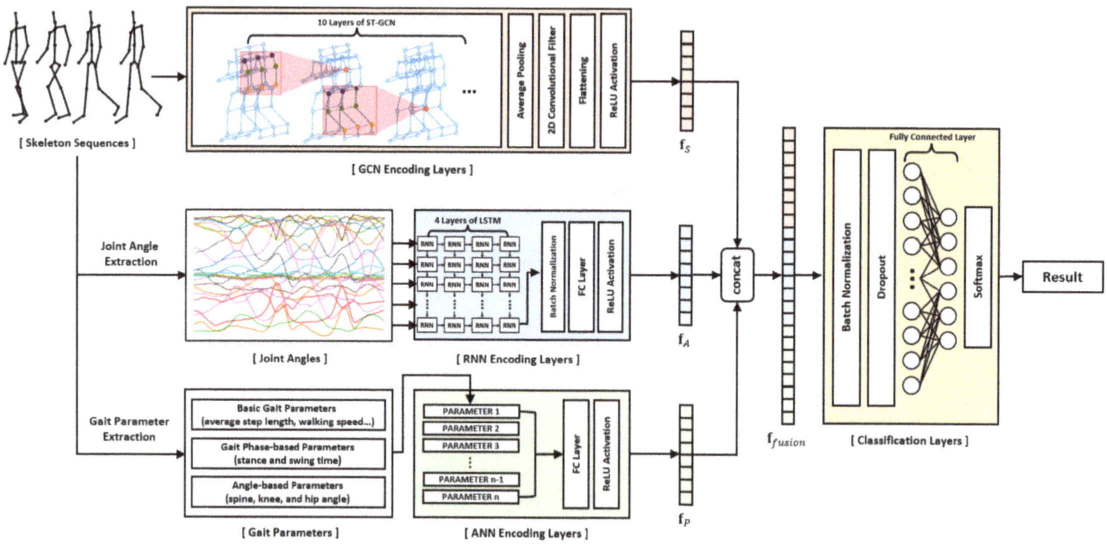

Figure 1. Structure of the proposed multi-input hybrid deep neural network. The skeleton sequences, joint angles, and gait parameters are input to the GCN, RNN, and ANN layers, respectively. Each encoding layer encodes the input data into a one-dimensional feature vector. The outputs of each encoding layer are concatenated together and fed to the final classification layer.

2.1. Data Acquisition

A depth camera-based skeleton data collection system was developed in the healthcare robotics laboratory at the Gwangju Institute of Science and Technology, Korea. An Azure Kinect and the corresponding body tracking software development kit (SDK) developed by Microsoft were used to collect the skeleton data. The system collected the data while a subject walked straight forward toward the sensor approximately 4 m away. The sensor was calibrated by recognizing an ArUco marker [33] to collect the data in the same coordinate system. The XYZ coordinate system of the sensor was transformed to the XYZ coordinate system of the marker. The 3-dimensional position of each vertex of the marker was measured using the sensor, and the transformation matrix was obtained. Each data example contained 80–120 frames of skeleton data with an average collection rate of 22.7 fps. We evaluated the proposed hybrid model on two skeleton datasets (a simulated pathological gait dataset and a vestibular disorder gait dataset) collected by the data acquisition system. The evaluation encompassed the assessment of the model's proficiency in handling intricate gait classification tasks, specifically its ability to differentiate among six distinct gait patterns within the simulated pathological gait dataset. Simultaneously, the vestibular disorder gait dataset, comprising genuine patient data, facilitated an examination of the model's practicality and suitability in real world contexts, particularly when confronted with datasets originating from individuals afflicted by vestibular disorders. This comprehensive evaluation allowed us to assess both the model's technical capabilities and its real-world applicability.

2.1.1. Simulated Pathological Gait Dataset

Most previous studies conducted binary classification by differentiating pathological gaits from a healthy gait [11,22–25], and there have also been a few studies recognizing various and complicated pathological gaits [3,17,27]. Multilabel pathological gait classification is much more difficult than binary classification and can help to evaluate a model from various directions. Therefore, we collected various and complicated pathological gait data. Normal gait and five pathological gaits, i.e., antalgic, steppage, lurching, stiff-legged, and Trendelenburg gaits, were collected through simulations of 12 healthy subjects. They were asked to walk along a 7 m walkway, as shown in Figure 2a. The characteristics and causes of each pathological gait are described in detail in [20]. The subjects understood the mechanical reason for the pathological gaits before data collection, so they could simulate the pathological gaits similarly to real patients. The data collection was conducted under strict expert supervision. The subjects were asked to walk with each gait 20 times. Therefore, 1,440 examples (12 subjects $\times$ 6 gaits $\times$ 20 walks) were included in this dataset. Furthermore, we augmented the dataset by reversing the left and right sides of the skeleton, so 2,880 examples were used for the experiments. Through this dataset, the performance of the proposed model on the classification of complicated pathological gaits was evaluated.

2.1.2. Vestibular Disorder Gait Dataset

Regardless of how well the simulated gait data were classified, it was difficult to verify their practicality in the real world. Therefore, we collected real patient data to evaluate the practical applicability of the proposed model. Gait data of real patients with vestibular problems were obtained with the support of the Kosin University Gospel Hospital. The subjects were asked to walk two laps around a 16 m track. We collected the skeleton data while the subjects were walking on the data collecting area, as shown in Figure 2b, because the sensor and the body tracking SDK do not guarantee high-quality skeleton data when the human is not facing the sensor. Thirty-three healthy subjects (12 females and 21 males, with a mean age of 38.9 $\pm$ 16.4 years) and 128 patients with a vestibular disorder (94 females and 34 males, with a mean age of 58.5 $\pm$ 13.5 years) participated in the data collection. Since there was a large difference in the average age between the healthy subjects and the patients, we downsampled the patient data and made a balanced group with 33 healthy subjects and 54 patients with a vestibular disorder (34 females and 20 males, with a mean age of 46.1 $\pm$ 10.4 years). Ten data examples whose sequences were less than 90 were excluded from the evaluation. Therefore, we evaluated the proposed model on the all-subject group with 312 data examples (161 subjects $\times$ 2 walks—10 exemptions) and the balanced group with 170 data examples (87 subjects $\times$ 2 walks—4 exemptions).

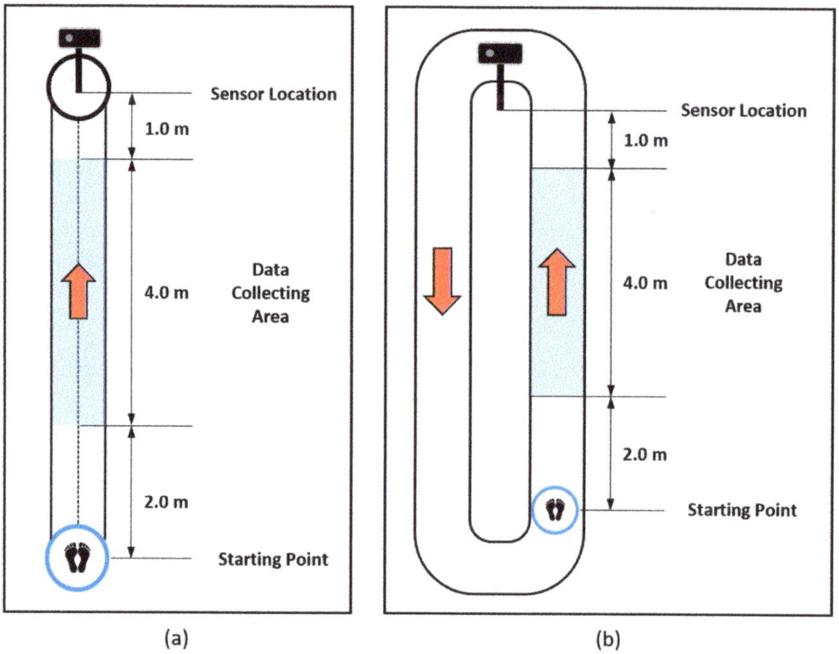

Figure 2. Data acquisition environment: (**a**) simulated pathological gait dataset and (**b**) vestibular disorder gait dataset.

2.2. Graph Convolutional Network for Skeleton Data

CNNs are renowned for their effectiveness in tasks involving visual data analysis, primarily due to their ability to capture intricate spatial relationships between pixels within an image, a feature that sets them apart [34]. In contrast, a GCN is particularly well suited for tasks such as node classification and link prediction within data structured as graphs [35]. Examples of such data encompass social networks, chemical molecules, and skeletal datasets. A GCN is known as the most powerful structure for skeleton-based action recognition. Yan et al. [36] first introduced a method to apply a spatial–temporal graph convolutional network (ST-GCN) for skeleton-based action recognition. They suggested a way to efficiently process skeleton sequences by simultaneously understanding the spatial and temporal characteristics of the skeleton data. Subsequently, many modified GCN structures have been introduced, and the performance of skeleton-based action recognition continues to improve [37–43]. In this study, we adopt the ideas and formulations of the ST-GCN proposed in [36] to encode skeleton data.

The skeleton sequences are denoted as a spatial–temporal graph $G = (V, E)$. The node set $V = \{v_{ti} | t = 1, \ldots, T, i = 1, \ldots, N\}$ contains all the joints in the skeleton sequences, where T and N denote the number of sequences and the number of joints, respectively. Every node includes three channels $v_{ti} = (x_{ti}, y_{ti}, z_{ti})$ since we use the 3-dimensional position information of each joint. The edge set E is divided into two subsets, the edge set of naturally connected human joints (intraskeleton edges) and the edge set of consecutive frames on the same joint (interframe edges), which are denoted by $E_S = \{v_{ti}v_{tj} | (i,j) \in H\}$ and $E_F = \{v_{ti}v_{(t+1)i}\}$, respectively, where H is the set of naturally connected joints.

The spatial convolution operation for a joint node v_{ti} can be formulated as the following equation:

$$F_{out}(v_{ti}) = \sum_{v_{tj} \in B(v_{ti})} \frac{1}{Z_{ti}(v_{tj})} F_{in}(\mathbf{p}(v_{ti}, v_{tj})) \cdot \mathbf{w}(M_{ti}(v_{tj})) \tag{1}$$

where F_{out} and F_{in} denote the output and input features of the GCN, respectively. $Z_{ti}(v_{tj})$ denotes a normalization term to balance the contribution of each subset. A sampling function $\mathbf{p}(v_{ti}, v_{tj})$ is defined on the neighbor set $B(v_{ti}) = \{v_{tj} | d(v_{tj}, v_{ti}) \leq D\}$, where $d(v_{tj}, v_{ti})$ denotes the minimum length of the path from v_{tj} to v_{ti}. A weight function $\mathbf{w}$ is defined by partitioning the neighbor set into subsets with a numeric label based on a mapping function $M_{ti}(v_{tj})$ that maps the neighbor nodes into their subset labels.

Yan et al. [36] extended the concept of a neighborhood to cover temporally consecutive joints by modifying the neighbor set $B(v_{ti})$ and defining a spatial–temporal mapping function M_{ST} as follows:

$$B(v_{ti}) = \{ v_{qj} | d(v_{tj}, v_{ti}) \leq K, |q - t| \leq \lfloor \Gamma/2 \rfloor \} \quad (2)$$

$$M_{ST}(v_{qj}) = M_{ti}(v_{tj}) + (q - t + \lfloor \Gamma/2 \rfloor) \times K \quad (3)$$

where K and Γ denote the number of subsets derived by mapping function $M_{ti}(v_{tj})$ and the range of interest for temporal convolution, respectively.

In this study, we implemented a multilayer ST-GCN to encode the skeleton sequences into a 1-dimensional feature vector $\mathbf{f}_S$. Global pooling was applied to the outputs of the ST-GCN layers, and then a convolutional operation was conducted to extract the feature vector with a specific size. Finally, the multichannel output was resized to a 1-dimensional vector.

2.3. Recurrent Neural Network for Joint Angles

We extracted the joint angle sequences from the skeleton sequences and used them as another input to the proposed hybrid model. We extracted the bending angles and link angles of specific joints according to [17], which showed their effectiveness on pathological gait recognition. Examples of the joint angles are shown in Figure 3.

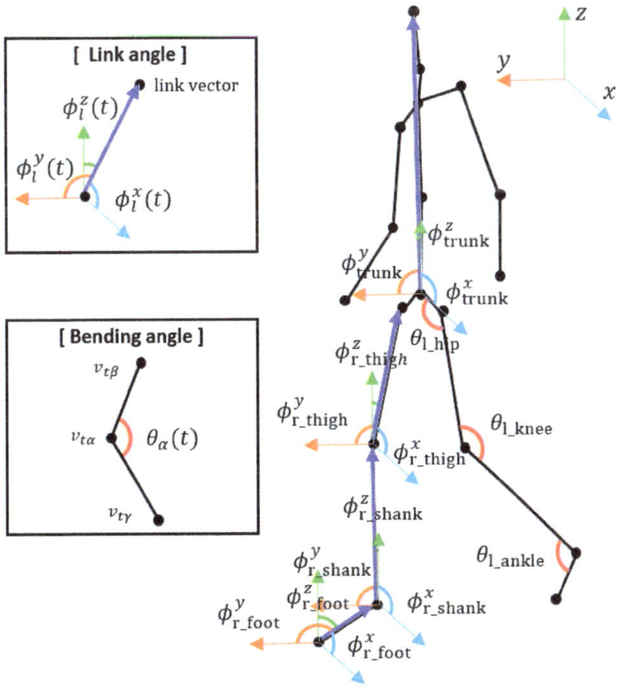

Figure 3. A sample of the joint angles. The joint angles include the link angles and the bending angles.

The bending angle of joint α ∈{left_hip, right_hip, left_knee, right_knee, left_ankle, right_ankle} at time t was calculated according to the following equation:

$$\theta_\alpha(t) = cos^{-1}\left(\frac{(v_{t\beta} - v_{t\alpha})^2 + (v_{t\gamma} - v_{t\alpha})^2 - (v_{t\gamma} - v_{t\beta})^2}{2(v_{t\beta} - v_{t\alpha}) \cdot (v_{t\gamma} - v_{t\alpha})}\right) \quad (4)$$

where β and γ denote the joints connected to joint α. For example, $\theta_{left\_knee}(t)$ was calculated using the left knee, left hip, and left ankle joints.

The link angle of link l ∈{left_thigh, right_thigh, left_shank, right_shank, left_foot, right_foot, trunk} about x-axis at time t was calculated using the following equation:

$$\phi_l^x(t) = cos^{-1}\left(\frac{(v_{tm} - v_{tn}) \cdot u_x}{|v_{tm} - v_{tn}||u_x|}\right) \quad (5)$$

where m and n denote the joints used to construct link l. The thigh consisted of the knee and hip joints, the shank consisted of the ankle and knee joints, the foot consisted of the tiptoe and ankle joints, and the trunk consisted of the pelvis and head joints. u_x denotes the unit vector along $x-$axis. $\phi_l^y(t)$ and $\phi_l^z(t)$ were calculated similarly by utilizing u_y and u_z instead of u_x.

A total of 25 angles were extracted from each skeleton and the sequences of the angles were fed to the RNN encoding layers. An RNN is a specialized architecture for handling sequential data, including time series data like stock prices, audio data, and skeletal sequences. Since the structure of a basic RNN has a long-term dependency problem in which the influence of the previous information continues to decrease as the hidden state is updated for long-term sequential data, we adopt LSTM to construct the RNN layers. LSTM can solve the problem by employing a gated structure to update the hidden state h_t. The variables for the gated structure, i.e., the forget gate f_t, input gate i_t, output gate o_t, and cell state C_t, are formulated as follows:

$$f_t = \sigma\left(W_{xf}x_t + W_{hf}h_{t-1} + b_f\right) \quad (6)$$

$$i_t = \sigma(W_{xi}x_t + W_{hi}h_{t-1} + b_i) \quad (7)$$

$$o_t = \sigma(W_{xo}x_t + W_{ho}h_{t-1} + b_o) \quad (8)$$

$$C_t = f_t \circ C_{t-1} + i_t \circ \tanh(W_{xC}x_t + W_{hC}h_{t-1} + b_C) \quad (9)$$

where x, W, b, σ, and $\circ$ denote the input, weights, biases, sigmoid function, and elementwise product, respectively. The hidden state h_t and the output y_t can be updated as follows:

$$h_t = o_t \circ \tanh(C_t) \quad (10)$$

$$y_t = W_y \cdot h_t + b_y. \quad (11)$$

We constructed a multilayer LSTM to encode the joint angle sequences into a 1-dimensional feature vector $\mathbf{f}_A$. The output of the multilayer LSTM operation was the last hidden state of the final LSTM layer, which was fed to the fully connected layer to extract a feature vector with a specific size as follows:

$$\mathbf{f}_A = \text{ReLU}(\mathbf{W}_A \text{LSTM}(\mathbf{x}_A) + \mathbf{b}_A) \quad (12)$$

where $\mathbf{x}_A$, $\mathbf{W}_A$, $\mathbf{b}_A$, and LSTM(·) denote the input joint angle sequences, weight, bias, and multilayer LSTM operation, respectively.

2.4. Artificial Neural Network for Gait Parameters

Gait parameters are important indicators to recognize pathological gaits [44–46]. We obtained basic gait parameters, gait phase-based parameters, and angle-based parameters using 3-dimensional skeleton sequences. These parameters encompass various aspects of walking and are instrumental in assessing an individual's gait.

The basic gait parameters include average step length, step length asymmetry, step width, and walking speed. Average step length measures the typical distance a person covers with a single step, typically from one heel to the other throughout a complete walking cycle. Step length asymmetry highlights the difference in step lengths between the left and right legs during walking, providing insights into the symmetry and balance of steps. Step width assesses the lateral distance between the feet at their widest point during the gait cycle, indicating whether steps are wide or narrow. Walking speed represents the rate of forward movement during walking, offering information about walking pace.

The gait phase-based parameters encompass stance and swing time on both legs. Swing time on both legs indicates the duration when the leg is not in contact with the ground during the gait cycle, typically measured in seconds, covering the time from foot lift off to foot strike. Stance time on both legs measures the duration of the gait cycle when the leg is in contact with the ground, providing essential information about how long each leg supports the body's weight during walking.

Lastly, the angle-based parameters encompass mean, minimum, and maximum values for frontal spine angle, lateral spine angle, knee angle, and hip angle. Frontal spine angle measures how the spine deviates from the vertical plane when viewed from the front, particularly relevant in posture and gait analysis for detecting deviations in the frontal plane. Lateral spine angle quantifies spine deviation from the vertical plane when viewed from the side, providing insights into body alignment during walking, especially lateral deviations. Knee angle on both legs describes the angle formed at the knee joint between the thigh and the lower leg for both legs during different phases of the gait cycle, offering insights into knee joint flexion and extension. Hip angle on both legs measures the angle at the hip joint between the thigh and the pelvis on both sides of the body during walking, reflecting hip movement and positioning throughout the gait cycle.

It is important to note that all parameters, except angle-related ones, fall within the range of 0 to 2, while angle-related parameters could potentially range from 0 to 180 degrees. To ensure consistency for input into an artificial neural network, these angle-related parameters were normalized by dividing them by 100. This normalization process brings them into the same 0 to 2 range as the other parameters, improving the training stability and convergence of an artificial neural network.

We used the extracted gait parameters as the final input data to the proposed model. A fully connected ANN was used to encode the gait parameters into a 1-dimensional feature vector $\mathbf{f}_P$ with a specific size as follows:

$$\mathbf{f}_P = \mathrm{ReLU}(\mathbf{W}_P \mathbf{x}_P + \mathbf{b}_P) \tag{13}$$

where $\mathbf{x}_P$, $\mathbf{W}_P$, and $\mathbf{b}_P$ denote the input gait parameters, weight, and bias, respectively.

2.5. Fusion of Features and Classification

The skeleton sequences, joint angle sequences, and gait parameters were input to the GCN-based, RNN-based, and ANN-based layers, respectively. The features extracted from each layer were concatenated into a 1-dimensional vector, and the integrated features were fed to the final classification layer using the following equations:

$$\mathbf{f}_{fusion} = \mathrm{concatenate}(\mathbf{f}_S, \mathbf{f}_A, \mathbf{f}_P) \tag{14}$$

$$\mathbf{y} = \mathrm{softmax}\left(\mathbf{W}_\mathbf{y} \mathbf{f}_{fusion} + \mathbf{b}_\mathbf{y}\right) \tag{15}$$

where $\mathbf{y}$, $\mathbf{W_y}$, and $\mathbf{b_y}$ denote the output, weight, and bias of the fully connected layer to recognize the gait type. Batch normalization and dropout were conducted before the operation of the fully connected layer to prevent overfitting. The index of the maximum value in $\mathbf{y}$ is the recognized gait type.

A cross-entropy loss function was adopted to calculate the loss L_{CE}, and L2 regularizations were applied to avoid overfitting as follows:

$$L_{CE}(\mathbf{y}, \hat{\mathbf{y}}) = -\sum_{i=1}^{C} y_i \log(\hat{y}_i) + \frac{\lambda}{2}|\mathbf{W}|^2 \tag{16}$$

where λ and $\mathbf{W}$ are the regularization parameter and trainable weights, respectively. Table 1 provides an exhaustive delineation of the intricate configuration of the multi-input hybrid neural network.

Table 1. The detailed configuration of the hybrid deep neural network architecture.

Type	Layer	Configuration
GCN Encoding Layers	ST-GCN	In_channels = 3, out_channels = 64, kernel_size = (9,3)
	ST-GCN	In_channels = 64, out_channels = 64, kernel_size = (9,3)
	ST-GCN	In_channels = 64, out_channels = 64, kernel_size = (9,3)
	ST-GCN	In_channels = 64, out_channels = 64, kernel_size = (9,3)
	ST-GCN	In_channels = 64, out_channels = 128, kernel_size = (9,3)
	ST-GCN	In_channels = 128, out_channels = 128, kernel_size = (9,3)
	ST-GCN	In_channels = 128, out_channels = 128, kernel_size = (9,3)
	ST-GCN	In_channels = 128, out_channels = 256, kernel_size = (9,3)
	ST-GCN	In_channels = 256, out_channels = 256, kernel_size = (9,3)
	ST-GCN	In_channels = 256, out_channels = 256, kernel_size = (9,3)
	Average Pooling	/
	2D Convolution	In_channels = 256, out_channels = 64, kernel_size = 1
	Flatten	/
	ReLU Activation	Hidden unit = 64
RNN Encoding Layers	LSTM	Hidden unit = 128, return_sequences = True
	LSTM	Hidden unit = 128, return_sequences = True
	LSTM	Hidden unit = 128, return_sequences = True
	LSTM	Hidden unit = 128, return_sequences = False
	Batch Normalization	/
	Fully Connected Layer	Hidden unit = 16
	ReLU Activation	/
ANN Encoding Layers	Fully Connected Layer	Hidden unit = 16
	ReLU Activation	/
Classification Layers	Batch Normalization	/
	Dropout	Ratio = 0.5
	Fully Connected Layer	Hidden unit = the number of classes
	Softmax	/

2.6. Training Environment

The experimental configurations of the computer are an Intel®Core™ i7-7700K central processing unit, an NVIDIA GeForce RTX 2080 Ti graphics processing unit, and 64 GB of random access memory. In this study, PyTorch and scikit-learn were adopted to implement the deep learning and machine learning models, respectively. All deep learning models used in the experiments were trained under the same training options (a batch size of 50, 200 training epochs, early stopping, cross-entropy loss, and the Adam [47] optimizer). However, some training options, such as the learning rate and weight decay, were set according to the suggested training configurations of each state-of-the-art model.

3. Results

The performance of the proposed model for skeleton-based pathological gait recognition was evaluated on the simulated pathological gait dataset and the vestibular disorder gait dataset. They include different gait abnormalities and have different subject configurations, so it is meaningful to evaluate the proposed model using both datasets and to compare the results. Furthermore, diverse state-of-the-art models for skeleton-based action recognition, such as the ST-GCN [36], hierarchical cooccurrence network (HCN) [37], decoupling GCN [38], two-stream adaptive graph convolutional network (2s-AGCN) [39], multistream attention-enhanced adaptive graph convolutional network (MS-AAGCN) [40], part-based graph convolutional network (PB-GCN) [41], decoupled spatial–temporal attention network (DSTA-NET) [42], and channelwise topology refinement graph convolutional network (CTR-GCN) [43], were compared to demonstrate the effectiveness of the proposed hybrid model.

3.1. Evaluation on the Simulated Pathological Gait Dataset

Leave-one-subject-out cross validation was applied to the simulated pathological gait dataset to compensate for the small number of subjects. The number of skeleton and joint angle sequences used as the input data was set to 100. We abandoned the last 10 sequences and used the 100 sequences immediately preceding them because the skeleton data were noisy if the distance between the human and the depth camera was too close.

We compared the performances of various models when only a single type of input data was used, as shown in Table 2. Classic machine learning-based models, such as AdaBoost, a decision tree, Gaussian Naïve Bayes, random forest, k-nearest neighbor (k-NN), and SVM, and deep neural network-based models, such as a multilayer perceptron (MLP), GRU [20], LSTM [20], and ST-GCN [36], were used for the comparison. For the gait parameters, the MLP achieved the best performance with 90.49% accuracy, and the SVM showed the second highest accuracy with 88.47% accuracy. For the joint angles, the RNN architectures showed their powerfulness in analysis. LSTM [20] achieved the best performance with 93.30% accuracy, and the GRU [20] showed the second highest accuracy with 92.92% accuracy. For the skeleton data, the ST-GCN [36] achieved the best performance with 96.94% accuracy, and the GRU [20] showed the second highest accuracy with 95.83% accuracy. Among the three input data types, the highest accuracy was achieved when the skeleton data were input to the ST-GCN [36] model.

Table 2. Accuracy of single input models on the simulated pathological gait dataset.

Model	Input Data		
	Gait Parameters (%)	**Joint Angles (%)**	**Skeleton (%)**
AdaBoost	76.20	56.70	55.24
decision tree	74.54	57.05	55.28
Gaussian Naïve Bayes	83.61	79.86	58.19
random forest	87.18	82.26	67.67
k-NN	83.98	76.46	74.44
SVM	88.47	78.82	72.64
MLP	**90.49**	79.97	86.11
GRU [20]	/	92.92	95.83
LSTM [20]	/	**93.30**	95.20
ST-GCN [36]	/	/	**96.94**

The proposed hybrid model fed with multiple types of input data achieved 99.03% accuracy, which was higher than that of the best single input models for each data type. The fusion of the gait features and the raw skeleton sequences improved the performance of pathological gait classification. Figure 4 shows the confusion matrices of the results of the single input models and the proposed hybrid model. The parameter-based classification had poor performance in classifying normal and Trendelenburg gaits, which were misclassified 115 and 89 times, respectively. The joint angle-based classification showed better performance than the parameter-based classification for this dataset. However, it showed poor performance in classifying the Trendelenburg gait with 74 misclassifications. The skeleton-based classification showed the best performance among the single input models. The overall gaits were well classified, but the antalgic gait classification seemed to need further improvement. The proposed multi-input hybrid model achieved the best performance by classifying the normal and five pathological gaits with a few errors.

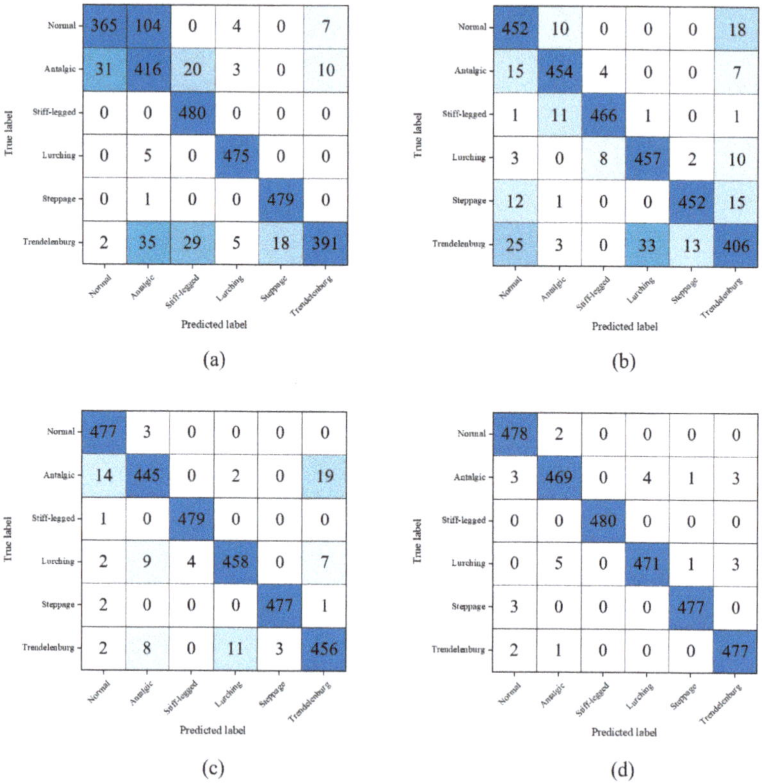

Figure 4. Confusion matrices of the best single input models and the proposed multi-input model on the simulated pathological gait dataset: (**a**) the gait parameters were input to the MLP model; (**b**) the joint angle sequences were input to the LSTM [20] model; (**c**) the skeleton sequences were input to the ST-GCN [36] model; and (**d**) all types of inputs were input to the proposed model.

The proposed model showed the highest accuracy among the state-of-the-art models for skeleton-based action recognition, as shown in Table 3. The CTR-GCN [43] and 2s-AGCN [39] showed the second and the third most accurate performances with 98.75% and 98.06% accuracy, respectively. The GCN-based models showed better performance than the RNN-based models. Although the state-of-the-art GCN models showed their powerfulness in skeleton-based pathological gait recognition, they could not surpass the performance of the proposed hybrid model combining the skeleton data and gait features.

Table 3. Accuracy of the state-of-the-art models and the proposed model on the simulated pathological gait dataset.

Model	Accuracy (%)
LSTM [20]	95.20
GRU [20]	95.83
ST-GCN [36]	96.94
HCN [37]	96.07

Table 3. *Cont.*

Model	Accuracy (%)
Decouple GCN [38]	96.25
2s-AGCN [39]	98.06
MS-AAGCN [40]	97.78
PB-GCN [41]	97.53
DSTA-NET [42]	97.91
CTR-GCN [43]	98.75
Proposed	**99.03**

3.2. Evaluation on the Vestibular Disorder Gait Dataset

We evaluated the proposed model using a real patient dataset to verify its practical applicability to the real world. Fivefold cross validation was applied for the evaluation of the vestibular disorder gait dataset since it had a large number of subjects. The number of skeleton and joint angle sequences used as the input data was set to 80. As before, we abandoned the final 10 sequences and used the 80 sequences immediately preceding them.

The proposed model was compared with state-of-the-art models for skeleton-based action recognition. The statistical indices (accuracy, sensitivity, specificity, and precision) of the models were evaluated on the all-subject group and the balanced group. Table 4 shows the accuracy, sensitivity, specificity, and precision for the all-subject group with the true positive (TP), false positive (FP), true negative (TN), and false negative (FN) values. The proposed model achieved 91.03% accuracy, 93.15% sensitivity, 82.81% specificity, and 95.45% precision. The accuracy and sensitivity of the proposed model were the highest among the models. MS-AAGCN [40] and CTR-GCN [43] showed the second and third highest accuracies of 89.74% and 89.42%, respectively. DSTA-NET [42] achieved the second highest sensitivity of 91.94%. The specificity and precision of the proposed model were not the highest among the models. The MS-AAGCN [40] and CTR-GCN [43] showed higher specificity than the proposed model by achieving 85.94%. The MS-AAGCN [40], CTR-GCN [43], and 2s-AGCN [39] showed higher precision than the proposed model by achieving 96.15%, 96.14%, and 95.61%, respectively. In the case of HCN, we excluded it from the comparison as it demonstrated an accuracy of less than 70% on this dataset.

Table 4. Performance on the all-subject group of the vestibular disorder gait dataset.

Model	All-Subject Group							
	Accuracy (%)	Sensitivity (%)	Specificity (%)	Precision (%)	TP	FP	TN	FN
LSTM [20]	84.94	87.50	75.00	93.13	217	16	48	31
GRU [20]	85.26	87.10	78.13	93.91	216	14	50	32
ST-GCN [36]	85.90	88.31	76.56	93.59	219	15	49	29
Decouple GCN [38]	86.22	87.10	82.81	95.15	216	11	53	32
2s-AGCN [39]	87.18	87.90	84.38	95.61	218	10	54	30
MS-AAGCN [40]	89.74	90.73	**85.94**	**96.15**	225	9	55	23
PB-GCN [41]	85.26	89.52	68.75	91.74	222	20	44	26
DSTA-NET [42]	86.86	91.94	67.19	91.57	228	21	43	20
CTR-GCN [43]	89.42	90.32	**85.94**	96.14	224	9	55	24
Proposed	**91.03**	**93.15**	82.81	95.45	231	11	53	17

Table 5 shows the results for the balanced group. The proposed model achieved 90.59% accuracy, 91.51% sensitivity, 89.06% specificity, and 93.27% precision. Similar to the results for the all-subject group, the accuracy and sensitivity of the proposed model were the highest among the models. The CTR-GCN [43] showed the second highest accuracy with 89.41%. DSTA-NET [42] achieved the second highest sensitivity of 90.57%. The specificity and precision of the proposed model were not the highest among the models. The CTR-GCN [43] and LSTM [20] showed higher specificity than the proposed model by achieving

92.19% and 90.63%, respectively. The CTR-GCN [43] and LSTM [20] also showed higher precision than the proposed model by achieving 94.90% and 93.33%, respectively.

Table 5. Performance on the balanced group of the vestibular disorder gait dataset.

Model	Balanced Group							
	Accuracy (%)	Sensitivity (%)	Specificity (%)	Precision (%)	TP	FP	TN	FN
LSTM [20]	83.53	79.25	90.63	93.33	84	6	58	22
GRU [20]	84.12	83.96	84.38	89.90	89	10	54	17
ST-GCN [36]	85.88	85.85	85.94	91.00	91	9	55	15
Decouple GCN [38]	86.47	86.79	85.94	91.09	92	9	55	14
2s-AGCN [39]	87.06	87.74	85.94	91.18	93	9	55	13
MS-AAGCN [40]	88.24	88.68	87.50	92.16	94	8	56	12
PB-GCN [41]	88.24	87.74	89.06	93.00	93	7	57	13
DSTA-NET [42]	85.88	90.57	78.13	87.27	96	14	50	10
CTR-GCN [43]	89.41	87.74	**92.19**	**94.90**	93	5	59	13
Proposed	**90.59**	**91.51**	89.06	93.27	97	7	57	9

We conducted an additional experiment to verify the effectiveness of using all of the gait parameters, joint angles, and skeleton data together. The encoding layers for the unused input data types were deactivated during the training, so the layers for only the used input data type affected the training. For example, the GCN layer was deactivated when the joint angles and the gait parameters were input to the model and the skeleton data were not used. The data of the balanced group were used for the evaluation, and the results are shown in Table 6. When a single type of input was fed to the model, using the skeleton data showed the best performance with 85.88% accuracy, 85.85% sensitivity, 85.94% specificity, and 91.00% precision. Using the gait parameters showed the second highest accuracy and sensitivity, with 79.41% and 80.19%, respectively. The results of using the joint angles showed the lowest accuracy and sensitivity of 72.35% and 64.15%, respectively. However, they showed higher specificity and precision than the results of using the gait parameters. When two types of input were fed to the model, the performance was improved compared with the results using the single type of input. Using the gait parameters and the skeleton data together showed the highest accuracy of 88.24%, which was the same as using the joint angles and the skeleton data together. Using the gait parameters and joint angles showed the lowest accuracy of 84.12%, which was 1.76% lower than the accuracy of using only the skeleton data. The highest sensitivity of 92.45% was achieved when using the joint angles and the skeleton data. The highest specificity and precision of 85.94% and 91.35%, respectively, were achieved when using the gait parameters and the skeleton data. When all types of input data were fed to the model, the highest accuracy, specificity, and precision were achieved compared to the results of using one or two types of input data. The sensitivity was lower than the results of using the joint angles and skeleton data.

Table 6. Performance of the proposed model on the balanced group of the vestibular disorder gait dataset when changing the combination of the inputs.

Number of Input Types	Input Type	Accuracy (%)	Sensitivity (%)	Specificity (%)	Precision (%)
1	Gait parameters	79.41	80.19	78.13	85.86
	Joint angles	72.35	64.15	85.94	88.31
	Skeleton	85.88	85.85	85.94	91.00
2	Gait parameters, joint angles	84.12	84.91	82.81	89.11
	Gait parameters, skeleton	88.24	89.62	85.94	91.35
	Joint angles, skeleton	88.24	**92.45**	81.25	89.09
3	Gait parameters, joint angles, skeleton	**90.59**	91.51	**89.06**	**93.27**

4. Discussion

4.1. Principal Findings

We have uncovered several principal findings in the evaluation of our proposed model for skeleton-based pathological gait recognition through our experiments. We conducted assessments on both the simulated pathological gait dataset and the vestibular disorder gait dataset, each characterized by distinct gait abnormalities and subject configurations. This dual evaluation approach was crucial to assess the model's robustness and versatility. The following findings promise to be a valuable contribution to the field of skeleton-based pathological gait recognition.

4.1.1. Effectiveness of Integration of Gait Parameters, Joint Angles, and Skeleton Data

This study first tried to use all of the gait parameters, joint angles, and skeleton data together for pathological gait recognition. GCN, RNN, and ANN layers were used to effectively encode the skeleton sequences, joint angle sequences, and gait parameters, respectively. The model showed the best stable performance among the state-of-the-art models for skeleton-based action recognition on the different datasets. The skeleton data contain all information, so the maximum performance could theoretically be achieved by using the skeleton data alone. However, it is not easy for a model to understand the characteristics of pathological gaits using sequential skeleton data since the data are large and complicated. The gait features can be key to making a model better understand the skeleton data and to improve the performance. The gait parameters and joint angles are manually extracted features whose performances have been verified through various studies. If they are input to the model together with the skeleton data, they can help the model understand the skeleton data in a more effective way since they are the essence of human knowledge for pathological gait recognition.

4.1.2. Performance Variation of Machine Learning Algorithms Based on Input Data Type

The machine learning algorithms were powerful when interpreting the gait parameters compared to the joint angles and skeleton data. Based on the results shown in Table 2, the average accuracies of the results of all machine learning algorithms for the gait parameters, joint angles, and skeleton data inputs were 82.33%, 71.86%, and 63.91%, respectively. The larger the volume of information was, the lower the performance of a machine learning-based classifier was. If the raw data were compressed to the features while preserving the important factors, the machine learning algorithms could better understand the distinguishable characteristics of the data and achieve better classification performance. However, there might be a loss of important factors when extracting features. Although the extracted features were powerful for the machine learning algorithms, the performance was lower than the results of feeding the raw skeleton sequences to the neural network-based classifiers.

4.1.3. Differential Effects of Joint Angles and Gait Parameters Depending on Dataset

The joint angles were more effective than the gait parameters for the simulated pathological gait dataset, as shown in Table 2. However, contradictory results were obtained for the vestibular disorder gait dataset, as shown in Table 6. The effectiveness of the gait parameters and the joint angles depends on the pathological gait type to be recognized. The joint angle sequences could be more effective for pathological gaits related to motor functionalities, such as joint bending, muscle compensation, and postural balancing. Gait parameters might be more effective for pathological gaits, such as Parkinson's disease and vestibular disorder gaits, which are related to sensory or cognitive functions and show irregular and unstable motions. Therefore, it is reasonable to use the gait parameters and joint angles together for the recognition of various gait abnormalities.

4.2. Comparison with Prior Work

The performances of existing studies for skeleton-based pathological gait recognition are as follows. Li et al. [3] classified normal controls, patients with hemiplegia, and patients with Parkinson's disease with 79.0% accuracy using a k-NN classifier. They achieved high accuracy even in noisy and low-resolution data without calibration or synchronization requirements. Dranca et al. [4] recognized the stages of Parkinson's disease with 93.40% accuracy using a Bayesian network with correlation-based feature selection. The key features for classification included left arm movement, trunk position during slightly displaced walking, and left shin angle for straight walking. An even higher accuracy of 96.23% was attained by focusing solely on features extracted from slightly displaced and spin walking steps. Seifallahi et al. [11] detected Alzheimer's disease with an accuracy of 92.31%, sensitivity of 96.33%, precision of 88.62%, and specificity of 90.81% using an SVM with a Gaussian kernel. Guo et al. [17] classified normal, in-toeing, out-toeing, drop-foot, pronation, and supination gaits with 90.75% accuracy using bidirectional LSTM. They integrated this algorithm into a mobile robot system with potential applications in assisting elderly or neurologically impaired patients at home to reduce fall risks and improve their quality of life. Chakraborty et al. [23] recognized cerebral palsy gaits with 98.59% accuracy using an SVM with a radial basis function kernel. According to the ReliefF feature ranking algorithm, the walking ratio was identified as the highest-ranked feature among classical gait features, and its inclusion in the classification process substantially enhanced the performance of all classifiers. Chakraborty et al. [24] recognized equinus foot deformity gaits with 88.3% accuracy using a multiple adaptive regression splines model. To enhance accuracy, they created feature vectors using six joint angles, encompassing hip, knee, and ankle angles on both sides, and integrated these vectors over multiple time instances.

Our proposed model distinguishes itself from existing research in the field of skeleton-based pathological gait recognition through the introduction of a novel methodology that adeptly integrates a diverse range of input data. In terms of our model's classification performance, it has exhibited exceptional proficiency in categorizing a broad spectrum of pathological gait types, encompassing normal, antalgic, steppage, lurching, stiff-legged, and Trendelenburg gaits, achieving an impressive accuracy rate of 99.03%. Furthermore, our model has demonstrated robust capabilities in discriminating vestibular disorder-related gaits, achieving accuracies of 91.03% in the all-subject group and 90.59% in the balanced group, confirming its applicability to real-world scenarios involving actual patients. However, it is essential to exercise caution when attempting to directly compare classification accuracies with previous studies within this context. This caution is warranted due to significant disparities in the pathological gait categories considered, the depth camera utilized for data acquisition, and the methodologies employed for pose estimation. Therefore, while our model's achieved accuracy is undeniably impressive within the specialized scope of our study, it is advisable to avoid making direct comparisons with earlier research, given the substantial divergence in pathological gait typologies and dataset characteristics.

We have been steadily studying skeleton-based pathological gait recognition. In our first study [18], we improved the performance of pathological gait recognition by applying an RNN autoencoder to extract features from raw skeleton sequences. The automatically extracted features were more effective than the raw skeleton sequences. However, it took a long time for training since the RNN autoencoder and classification model were trained separately. The walking gait dataset [48] used in the research was composed of simple abnormal gait patterns created by padding a sole or attaching a weight, which were relatively easy to classify. Furthermore, the skeleton data were collected on a treadmill, so the subjects might not generate a natural gait pattern. For these reasons, we needed to collect complicated pathological gait data using our Kinect system without a treadmill. Therefore, in our second study [20], we collected complicated pathological gait patterns, i.e., antalgic, steppage, lurching, stiff-legged, and Trendelenburg gaits. We also proposed a GRU-based end-to-end model and applied a joint selection strategy to increase the performance. In our third study [49], we added a foot pressure sensor to our Kinect system.

The performance of pathological gait recognition was further improved by using foot pressure and skeleton data together. However, the multimodal system was complicated, and the foot pressure sensor was costly, so it was not suitable for practical use. Therefore, we studied a method for maximizing pathological gait recognition performance using a single Kinect sensor and eventually arrived at the hybrid model proposed in this study.

4.3. Limitations and Future Works

This study had several limitations. First, the proposed hybrid model was validated on the limited types of pathological gaits. In the future, we should collect abnormal gait data with diverse diseases, such as Parkinson's disease, autism spectrum disorder, stroke, Alzheimer's disease, sarcopenia, and functional gait disorders, and validate the proposed model on those data to verify the application validity on other pathological gaits. Second, we focused on skeleton-based pathological gait recognition, so we only used skeleton-induced gait parameters, whereas there are many other clinical parameters, such as body composition analysis, hemanalysis, video head impulse test, videonystagmography, Montreal cognitive assessment, mini-mental state exam, time up and go test, or the Tinetti test. Therefore, in the future, we plan to collect those clinical parameters and apply them to the proposed model to improve the performance of pathological gait recognition. Third, we just concatenated the encoded features and fed them to the classification layer. In the future, we aim to further enhance our method by implementing adaptive feature selection or weighting techniques. This will involve the selection or assignment of weights to features and matching scores that demonstrate low correlation and high discrimination. Fourth, the ST-GCN, LSTM, and ANN are effective neural network architectures to encode the skeleton data, joint angle sequences, and gait parameters, respectively, but might not be the best neural network architectures to encode each of them. The proposed model can be further improved by replacing the encoding layers with advanced algorithms optimized for each input data type. In the future, we will continue to modify the current neural network architectures with the latest algorithm and reflect on it in future works.

5. Conclusions

The proposed hybrid deep neural network, which effectively used gait parameters, joint angles, and skeleton data, improved the performance of pathological gait recognition on two different datasets. The proposed model not only classified the diverse pathological gaits (simulated) but also recognized the gait abnormalities of real patients with a vestibular disorder. The fusion of the different inputs had a positive synergy on pathological gait recognition by integrating the features based on human knowledge and those automatically extracted by artificial intelligence. The framework can provide inspiration for the development of skeleton-based pathological gait recognition models. Furthermore, it can be flexibly modified by replacing the encoding layers or adding clinical information, which can further improve the performance. In the future, we will collect skeleton datasets for patients with other diseases, such as Parkinson's disease and sarcopenia, and evaluate the performance when classifying different diseases to verify the practical use and expansion of the application.

Author Contributions: Conceptualization, K.J. and M.S.K.; methodology, K.J.; software, K.J. and S.L.; data acquisition K.L. and H.L.; validation, K.J., K.L. and H.L.; formal analysis, K.J.; investigation, K.J., M.S.K. and H.L.; original draft preparation, K.J. and M.S.K.; writing—review and editing, K.J. and M.S.K.; visualization, K.J.; and supervision, M.S.K. and H.L. All authors have read and agreed to the published version of the manuscript.

Funding: This research was partly funded by the Open AI Dataset Project (AI-Hub, S. Korea), by Korea Health Industry Development Institute (KHIDI), Korea, under the project (2021-121), and by the Ministry of Trade, Industry, and Energy (MOTIE), Korea, under the project (P0024456) supervised by the Korea Institute for Advancement of Technology (KIAT).

Institutional Review Board Statement: The study was conducted in accordance with the Declaration of Helsinki and approved by the Institutional Review Board of the Kosin University Gospel Hospital (protocol code 2022-04-024 and date of approval 4 July 2022).

Informed Consent Statement: Not applicable.

Data Availability Statement: Not applicable.

Acknowledgments: The authors are grateful to all staff and participants involved in the data collection process.

Conflicts of Interest: The authors declare no conflict of interest.

References

1. Connor, P.; Ross, A. Biometric recognition by gait: A survey of modalities and features. *Comput. Vis. Image Underst.* **2018**, *167*, 1–27. [CrossRef]
2. Whittle, M.W. Clinical gait analysis: A review. *Hum. Mov. Sci.* **1996**, *15*, 369–387. [CrossRef]
3. Li, Q.; Wang, Y.; Sharf, A.; Cao, Y.; Tu, C.; Chen, B.; Yu, S. Classification of gait anomalies from Kinect. *Vis. Comput.* **2016**, *34*, 229–241. [CrossRef]
4. Dranca, L.; de Abetxuko Ruiz de Mendarozketa, L.; Goñi, A.; Illarramendi, A.; Navalpotro Gomez, I.; Delgado Alvarado, M.; Rodríguez-Oroz, M.C. Using Kinect to classify Parkinson's disease stages related to severity of gait impairment. *BMC Bioinf.* **2018**, *19*, 471. [CrossRef] [PubMed]
5. Alharthi, A.S.; Casson, A.J.; Ozanyan, K.B. Gait spatiotemporal signal analysis for Parkinson's disease detection and severity rating. *IEEE Sens. J* **2021**, *21*, 1838–1848. [CrossRef]
6. Alkhatib, R.; Diab, M.O.; Corbier, C.; Badaoui, M.E. Machine learning algorithm for gait analysis and classification on early detection of Parkinson. *IEEE Sens. Lett* **2020**, *4*, 1–4. [CrossRef]
7. Hasan, C.Z.C.; Jailani, R.; Tahir, N.M.; Yassin, I.M.; Rizman, Z.I. Automated classification of autism spectrum disorders gait patterns using discriminant analysis based on kinematic and kinetic gait features. *J. Appl. Env. Biol. Sci.* **2017**, *7*, 150–156.
8. Hasan, C.Z.C.; Jailani, R.; Tahir, N.M.; Ilias, S. The analysis of three-dimensional ground reaction forces during gait in children with autism spectrum disorders. *Res. Devel Disabil.* **2017**, *66*, 55–63. [CrossRef]
9. Wang, M.; Yong, S.; He, C.; Chen, H.; Zhang, S.; Peng, C.; Wang, X.A. Research on abnormal gait recognition algorithms for stroke patients based on array pressure sensing system. In Proceedings of the IEEE 3rd Information Technology, Networking, Electronic and Automation Control Conference (ITNEC), Chengdu, China, 15–17 March 2019; pp. 1560–1563.
10. Park, S.J.; Hussain, I.; Hong, S.; Kim, D.; Park, H.; Benjamin, H.C.M. Real-time gait monitoring system for consumer stroke prediction service. In Proceedings of the IEEE International conference on consumer electronics (ICCE), Las Vegas, NV, USA, 4–6 January 2020; pp. 1–4.
11. Seifallahi, M.; Soltanizadeh, H.; Mehraban, A.H.; Khamseh, F. Alzheimer's disease detection using skeleton data recorded with Kinect camera. *Clust. Comput.* **2020**, *23*, 1469–1481. [CrossRef]
12. Marchetti, G.F.; Whitney, S.L.; Blatt, P.J.; Morris, L.O.; Vance, J.M. Temporal and spatial characteristics of gait during performance of the dynamic gait index in people with and people without balance or vestibular disorders. *Phys. Ther.* **2008**, *88*, 640–651. [CrossRef]
13. Schniepp, R.; Möhwald, K.; Wuehr, M. Clinical and automated gait analysis in patients with vestibular, cerebellar, and functional gait disorders: Perspectives and limitations. *J. Neurol.* **2019**, *266*, 118–122. [CrossRef] [PubMed]
14. Strupp, M.; Dlugaiczyk, J.; Ertl-Wagner, B.B.; Rujescu, D.; Westhofen, M.; Dieterich, M. Vestibular disorders: Diagnosis, new classification and treatment. *Dtsch. Ärzteblatt Int.* **2020**, *117*, 300.
15. Slijepcevic, D.; Zeppelzauer, M.; Schwab, C.; Raberger, A.M.; Breiteneder, C.; Horsak, B. Input representations and classification strategies for automated human gait analysis. *Gait Posture* **2020**, *76*, 198–203. [CrossRef] [PubMed]
16. Slijepcevic, D.; Zeppelzauer, M.; Gorgas, A.M.; Schwab, C.; Schüller, M.; Baca, A.; Breiteneder, C.; Horsak, B. Automatic classification of functional gait disorders. *IEEE J. Biomed. Health Inform.* **2018**, *22*, 1653–1661. [CrossRef] [PubMed]
17. Guo, Y.; Deligianni, F.; Gu, X.; Yang, G.Z. 3-D canonical pose estimation and abnormal gait recognition with a single RGB-D camera. *IEEE Robot. Autom. Lett.* **2019**, *4*, 3617–3624. [CrossRef]
18. Jun, K.; Lee, D.W.; Lee, K.; Lee, S.; Kim, M.S. Feature extraction using an RNN autoencoder for skeleton-based abnormal gait recognition. *IEEE Access* **2020**, *8*, 19196–19207. [CrossRef]
19. Chen, F.; Cui, X.; Zhao, Z.; Zhang, D.; Ma, C.; Zhang, X.; Liao, H. Gait acquisition and analysis system for osteoarthritis based on hybrid prediction model. *Comput. Med. Imaging Graph.* **2020**, *85*, 101782. [CrossRef]
20. Jun, K.; Lee, Y.; Lee, S.; Lee, D.W.; Kim, M.S. Pathological gait classification using Kinect v2 and gated recurrent neural networks. *IEEE Access* **2020**, *8*, 139881–139891. [CrossRef]

21. Lee, D.W.; Jun, K.; Lee, S.; Ko, J.K.; Kim, M.S. Abnormal gait recognition using 3D joint information of multiple Kinects system and RNN-LSTM. In Proceedings of the 41st Annual International Conference of the IEEE Engineering in Medicine and Biology Society (EMBC), Berlin, Germany, 23–27 July 2019; pp. 542–545.
22. Bei, S.; Zhen, Z.; Xing, Z.; Taocheng, L.; Qin, L. Movement disorder detection via adaptively fused gait analysis based on Kinect sensors. *IEEE Sens. J.* **2018**, *18*, 7305–7314. [CrossRef]
23. Chakraborty, S.; Nandy, A. Automatic diagnosis of cerebral palsy gait using computational intelligence techniques: A low-cost multi-sensor approach. *IEEE Trans. Neural Syst. Rehabil. Eng.* **2020**, *28*, 2488–2496. [CrossRef]
24. Chakraborty, S.; Jain, S.; Nandy, A.; Venture, G. Pathological gait detection based on multiple regression models using unobtrusive sensing technology. *J. Signal Process. Syst.* **2020**, *93*, 1–10. [CrossRef]
25. Tian, H.; Ma, X.; Wu, H.; Li, Y. Skeleton-based abnormal gait recognition with spatio-temporal attention enhanced gait-structural graph convolutional networks. *Neurocomputing* **2022**, *473*, 116–126. [CrossRef]
26. Sadeghzadehyazdi, N.; Batabyal, T.; Acton, S.T. Modeling spatiotemporal patterns of gait anomaly with a CNN-LSTM deep neural network. *Expert Syst. Appl.* **2021**, *185*, 115582. [CrossRef]
27. Kim, J.; Seo, H.; Naseem, M.T.; Lee, C.S. Pathological-gait recognition using spatiotemporal graph convolutional networks and attention model. *Sensors* **2022**, *22*, 4863. [CrossRef] [PubMed]
28. Liu, X.; You, Z.; He, Y.; Bi, S.; Wang, J. Symmetry-driven hyper feature GCN for skeleton-based gait recognition. *Pattern Recognit.* **2022**, *125*, 108520. [CrossRef]
29. Teepe, T.; Khan, A.; Gilg, J.; Herzog, F.; Hormann, S.; Rigoll, G. Gaitgraph: Graph convolutional network for skeleton-based gait recognition. In Proceedings of the IEEE International Conference on Image Processing (ICIP), Anchorage, AK, USA, 19–22 September 2021; pp. 2314–2318.
30. Mao, M.; Song, Y. Gait recognition based on 3D skeleton data and graph convolutional network. In Proceedings of the IEEE International Joint Conference on Biometrics (IJCB), Houston, TX, USA, 28 September–1 October 2020; pp. 1–8.
31. Leng, L.; Zhang, J. Palmhash code vs. palmphasor code. *Neurocomputing* **2013**, *108*, 1–12. [CrossRef]
32. Leng, L.; Li, M.; Kim, C.; Bi, X. Dual-source discrimination power analysis for multi-instance contactless palmprint recognition. *Multimed. Tools Appl.* **2017**, *76*, 333–354. [CrossRef]
33. Garrido-Jurado, S.; Muñoz-Salinas, R.; Madrid-Cuevas, F.J.; Marín-Jiménez, M.J. Automatic generation and detection of highly reliable fiducial markers under occlusion. *Pattern Recognit.* **2014**, *47*, 2280–2292. [CrossRef]
34. Alahmari, F.; Naim, A.; Alqahtani, H. E-Learning Modeling Technique and Convolution Neural Networks in Online Education. In *IoT-enabled Convolutional Neural Networks: Techniques and Applications*, 1st ed.; Naved, M., Devi, V.A., Gaur, L., Elngar, A.A., Eds.; River Publishers: New York, NY, USA, 2023; pp. 261–295.
35. Krichen, M. Convolutional Neural Networks: A Survey. *Computers* **2023**, *12*, 151. [CrossRef]
36. Yan, S.; Xiong, Y.; Lin, D. Spatial temporal graph convolutional networks for skeleton-based action recognition. In Proceedings of the AAAI Conference on Artificial Intelligence, New Orleans, LA, USA, 2–7 February 2018; pp. 7444–7452.
37. Li, C.; Zhong, Q.; Xie, D.; Pu, S. Co-occurrence feature learning from skeleton data for action recognition and detection with hierarchical aggregation. *arXiv* **2018**, arXiv:1804.06055.
38. Cheng, K.; Zhang, Y.; Cao, C.; Shi, L.; Cheng, J.; Lu, H. Decoupling GCN with dropgraph module for skeleton-based action recognition. In Proceedings of the Computer Vision—ECCV 2020: 16th European Conference, Glasgow, UK, 23–28 August 2020; pp. 536–553.
39. Shi, L.; Zhang, Y.; Cheng, J.; Lu, H. Two-stream adaptive graph convolutional networks for skeleton-based action recognition. In Proceedings of the IEEE/CVF Conf Comput Vision Pattern Recognition, Long Beach, CA, USA, 15–20 June 2019; pp. 12018–12027.
40. Shi, L.; Zhang, Y.; Cheng, J.; Lu, H. Skeleton-based action recognition with multi-stream adaptive graph convolutional networks. *IEEE Trans. Image Process.* **2020**, *29*, 9532–9545. [CrossRef] [PubMed]
41. Thakkar, K.; Narayanan, P.J. Part-based graph convolutional network for action recognition. *arXiv* **2018**, arXiv:1809.04983.
42. Shi, L.; Zhang, Y.; Cheng, J.; Lu, H. Decoupled spatial-temporal attention network for skeleton-based action-gesture recognition. In Proceedings of the Asian Conference on Computer Vision, Kyoto, Japan, 30 November–4 December 2020; pp. 38–53.
43. Chen, Y.; Zhang, Z.; Yuan, C.; Li, B.; Deng, Y.; Hu, W. Channel-wise topology refinement graph convolution for skeleton-based action recognition. In Proceedings of the IEEE/CVF International Conference On Computer Vision, Montreal, QC, Canada, 10–17 October 2021; pp. 13359–13368.
44. Zanardi, A.P.J.; da Silva, E.S.; Costa, R.R.; Passos-Monteiro, E.; Dos Santos, I.O.; Kruel, L.F.M.; Peyré-Tartaruga, L.A. Gait parameters of Parkinson's disease compared with healthy controls: A systematic review and meta-analysis. *Sci. Rep.* **2021**, *11*, 752. [CrossRef] [PubMed]
45. Rocha, P.A.; Porfírio, G.M.; Ferraz, H.B.; Trevisani, V.F.M. Effects of external cues on gait parameters of Parkinson's disease patients: A systematic review. *Clin. Neurol. Neurosurg.* **2014**, *124*, 127–134. [CrossRef] [PubMed]
46. Liu, K.; Uygur, M.; Kaminski, T.W. Effect of ankle instability on gait parameters: A systematic review. *Athl. Train. Sports Health Care* **2012**, *4*, 275–281. [CrossRef]
47. Kingma, D.P.; Ba, J. Adam: A method for stochastic optimization. *arXiv* **2014**, arXiv:1809.04983.

48. Nguyen, T.N.; Huynh, H.H.; Meunier, J. 3D reconstruction with time-of-flight depth camera and multiple mirrors. *IEEE Access* **2018**, *6*, 38106–38114. [CrossRef]
49. Jun, K.; Lee, S.; Lee, D.W.; Kim, M.S. Deep learning-based multimodal abnormal gait classification using a 3D skeleton and plantar foot pressure. *IEEE Access* **2021**, *9*, 161576–161589. [CrossRef]

bioengineering

MDPI

Article

PPChain: A Blockchain for Pandemic Prevention and Control Assisted by Federated Learning

Tianruo Cao, Yongqi Pan *, Honghui Chen, Jianming Zheng and Tao Hu

Science and Technology on Information Systems Engineering Laboratory, National University of Defense Technology, Changsha 410073, China
* Correspondence: pianyu_shiguang@163.com

Abstract: Taking COVID-19 as an example, we know that a pandemic can have a huge impact on normal human life and the economy. Meanwhile, the population flow between countries and regions is the main factor affecting the changes in a pandemic, which is determined by the airline network. Therefore, realizing the overall control of airports is an effective way to control a pandemic. However, this is restricted by the differences in prevention and control policies in different areas and privacy issues, such as how a patient's personal data from a medical center cannot be effectively combined with their passenger personal data. This prevents more precise airport control decisions from being made. To address this, this paper designed a novel data-sharing framework (i.e., PPChain) based on blockchain and federated learning. The experiment uses a CPU i7-12800HX and uses Docker to simulate multiple virtual nodes. The model is deployed to run on an NVIDIA GeForce GTX 3090Ti GPU. The experiment shows that the relationship between a pandemic and aircraft transport can be effectively explored by PPChain without sharing raw data. This approach does not require centralized trust and improves the security of the sharing process. The scheme can help formulate more scientific and rational prevention and control policies for the control of airports. Additionally, it can use aerial data to predict pandemics more accurately.

Keywords: blockchain; federated learning; pandemic prevention and control; privacy-preserving

Citation: Cao, T.; Pan, Y.; Chen, H.; Zheng, J.; Hu, T. PPChain: A Blockchain for Pandemic Prevention and Control Assisted by Federated Learning. *Bioengineering* **2023**, *10*, 965. https://doi.org/10.3390/bioengineering10080965

Academic Editors: Hongqing Yu, Alaa AlZoubi, Yifan Zhao and Hongbo Du

Received: 28 April 2023
Revised: 27 June 2023
Accepted: 3 July 2023
Published: 15 August 2023

1. Introduction

The outbreak of coronavirus disease 2019 (COVID-19) had a huge impact on the world economy and people's lives. It is in the interest of all mankind to contain a pandemic at an early date. However, the spread of the pandemic is affected by many factors. Pandemic prevention policies and pandemic prevention psychology in different countries will have different effects [1], thus having a significant impact on the model parameters of virus transmission. Raffetti et al. [2] demonstrated that national policies are the most important factor affecting the spread of the pandemic. Sweden adopted a "natural" herd-immunity strategy to deal with the pandemic, but this model resulted in a COVID-19 death rate 10 times higher in Sweden than that in neighboring Norway [3]. Mishra et al. [4] compared Denmark, Britain, and Sweden under different policy models. They concluded that domestic policy effects are affected by inter-country population flows and that similar policies may even produce different effects.

Therefore, it is necessary to include population flow between countries and regions when designing strategies for the prevention and control of a pandemic, as well as develop effective policies about population flow to contain the outbreak. This paper aims to study the impact of air network transmission on the pandemic and how to formulate shipping policies to contain the pandemic. However, at present, most national medical centers are often unable to know the flight status of confirmed patients, and most airline companies cannot obtain the illness status of passengers due to different national policies. Meanwhile, federated learning [5] can realize multi-party data integration without sharing data. In this paper, federated learning is chosen instead of other distributed learning frameworks.

The main reason is that the compute nodes have absolute control over the data in federated learning, and the central server cannot directly or indirectly operate the data of the compute nodes. The compute nodes can stop computing and communication at any time and exit the learning process. However, in other distributed machine-learning frameworks (such as MapReduce, etc.), the central server has a high level of control over the compute nodes and their data. The compute nodes are completely controlled by the central server and receive instructions from the central server. For instance, MapReduce's central server can issue instructions to the compute nodes to exchange data with each other. This can potentially compromise the privacy of user data and may add additional communication overhead. Meanwhile, the raw data of federated learning can be kept locally, which is an advantage over other distributed machine-learning approaches.

The traditional federated learning model often requires an aggregator. This will lead to privacy and model failure problems [6] if the center aggregator is attacked. Our work is dedicated to enabling federated learning in association with blockchain to engage safely with collaborative training without a central aggregator and apply it to pandemic prevention.

In this paper, we present the following: First, we design a pandemic prevention and control model based on a blockchain. The characteristics of blockchain such as decentralization and security ensure the reliability and effectiveness of the learning process. This causes federated learning to not depend on the third-party central server, but rather depend on the consensus mechanism for better data protection and security. The scheme implements the integration of pandemic data and airline data through federated learning. Second, the impact of aviation network transmission on pandemic prevention and control is obtained, thus providing support for scientific predictions of pandemic changes and auxiliary policy making.

The rest of this article is organized as follows: Section 2 investigates the current situation related to the use of federated learning for pandemic prevention and control. Section 3 mainly elaborates on the main architecture of the model. Section 4 introduces the experiment and analysis. Section 5 summarizes the main work of this paper.

2. Related Work and Prior Knowledge

2.1. Application of Federated Learning in Epidemic

Qian et al. [7] described real-world cases of using federated learning in COVID-19 as well as non-COVID-19 scenarios and analyzed its limitations and practical challenges. References [8–13] used federated learning to assist in the diagnosis and intelligent monitoring of COVID-19. However, the above schemes only improved the accuracy of the diagnosis of suspected patients but did not discuss how to evaluate the effect of the policy. Chen et al. [14] constructed a COVID-19 vulnerability prediction map using federated learning synergy to identify high-risk areas and reduce the spread of the disease. However, the method is mainly oriented toward the collaboration of organizations in the same field and does not provide a cross-domain collaboration approach. Samuel et al. [15] proposed a privacy architecture based on federated learning and blockchain technology to support the cross-domain interaction regarding COVID-19 information and protect the authenticity and privacy of this information. Pang et al. [16] fused urban digital twins between multiple cities through federated learning technology and constructed a collaborative urban crisis management paradigm to explore and formulate effective prevention and control policies. However, the method lacks the competence to analyze the impact of population movement between regions on pandemic prevention.

2.2. Blockchain Technology

In 2008, Satoshi Nakamoto published a white paper on Bitcoin [17], marking the birth and landing of blockchain technology. Blockchain technology integrates P2P networks, confidential algorithms, consensus mechanisms, and other technologies that can be used for secure and trusted data storage, transmission, and operation. Today, the blockchain has developed into the "blockchain 3.0" era [18], and it is integrated with various industries through smart contracts [19]. The essence of blockchain technology is a distributed shared ledger. In an environment of incomplete trust, a P2P network is built to verify and store data through a chain data structure, and a consensus mechanism and encryption algorithm are adopted to achieve trusted transmission and operation of data. A smart contract is an automatically executed association deployed on the blockchain based on established rules [20], which is another form of blockchain consensus. Once the condition is met, execution is triggered and cannot be terminated. Smart contracts have evolved from scripts to programming languages. Bitcoin's scripting support is limited to programming and is not Turing-complete. Ethereum [21] provides users with a Turing-complete programming language but requires users to learn additional specific languages. Hyperledger [22] supports users to write smart contracts in programming languages such as GO, and uses Docker containers as the running environment, lowering the development threshold for writing smart contracts. Basetty Mallikarjuna et al. [23] applied deep neural network (DNN) analysis in healthcare and the COVID-19 pandemic and presented smart contract procedures to identify feature data (FED) extracted from existing data. Shan Jiang et al. proposed BlocHIE [24], a Blockchain-based platform for healthcare information exchange. They also designed a bloom filter [25] to select a low-frequency keyword from the multiple keywords input by the ITS data owner.

2.3. *Application of Blockchain in Federated Learning*

Chen et al. [26] analyzed the privacy and security issues of the learning model and designed a federated learning system that supports privacy protection based on blockchain, replacing the central server with parameter aggregation. Ramanan et al. [27] proposed a federated learning environment based on blockchain which uses blockchain to store and share global models and perform model aggregation tasks through smart contracts. Rehman et al. [28] proposed a set of blockchain-based federated learning frameworks for mobile edge computing networks, redefining the model's storage mode, training process, and consensus mechanism. Anik Islam et al. [29] presented a federated learning-based data-accumulation scheme that combines drones and blockchain. Zhu et al. [30] comprehensively surveyed challenges, solutions, and future directions for blockchain-empowered federated learning.

To summarize, federated learning for pandemic analysis is currently mainly used to integrate knowledge in the field and relies on blockchain technology to strengthen the robustness and credibility of federal learning. However, the research on the interaction between regional policies is still insufficient. On the other hand, the storage overhead of a blockchain is large, which has a great impact on efficiency when federated learning requires more model parameters. Therefore, it is necessary to study how to integrate cross-domain information while protecting the rights and interests of data owners. Through this method, we aim to identify the impact of inter-regional population flow on pandemic prevention and control to better develop pandemic predictions and policy making.

3. PPChain

In this section, we propose a blockchain for COVID-19 prevention and control based on federated learning. With blockchain technology, data can be trusted and distributed among multiple parties without a third-party central authority, reducing the risk of data leakage. There are two main types of participants in the network: regional pandemic management centers and individual airlines. Using the federated learning technique, the two groups of entities can figure out how the disease is spreading through flights without sharing raw data, which allows for the development of targeted containment policies and predictions. The system uses smart contracts to act as aggregators. In addition, to ensure the efficiency of the blockchain network, the models are trained locally and the blockchain only utilizes parameter information flow. The system architecture is shown in Figure 1.

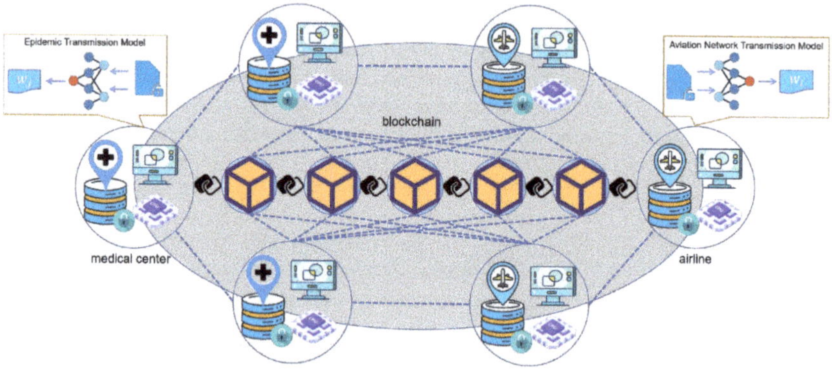

Figure 1. The system architecture.

According to the above analysis, the pandemic mainly spreads and spreads rapidly through the transportation network. Therefore, we choose the representative transportation route with the widest spread and fastest speed, namely the flight, to study the transmission mode of the pandemic, to better predict and prevent the pandemic. This can also be extended to multiple transport networks. The transmission of the pandemic should be coupled with the transportation network, but the traditional pandemic analysis and prediction models such as SIR Do not consider the factors of the transportation network. However, the current research on the effect of transportation networks on pandemic transmission does not have the support of pandemic transmission dynamics. Therefore, this paper hopes to train the SIR model in collaboration with the communication model of the traffic network, to better predict and prevent the pandemic. Specifically, federated learning enables collaborative learning of two different domain models without sharing their respective training data. In addition, the effect of traffic network changes can be directly transmitted to the change in epidemic model parameters without lag. Blockchain provides the building blocks for trusted collaboration and protects the security of training. The traditional combination of blockchain and federated learning is mostly collaborative training between the same models, and this paper is inclined to study collaborative training between different models.

Using the framework shown in Figure 1, the training process of federated learning is divided into two parts. First, the sample space is aligned based on people with the same identity information who are distributed to different parties. Using the user ID alignment technique based on encryption ensures that the original data of each party do not need to be exposed. In the second phase, the encryption model training is performed based on these aligned entities as follows:

- After identifying the shared entity, the blockchain triggers the smart contract to create the key pair and send the public key to regional pandemic management centers and individual airlines.
- Regional pandemic management centers and individual airlines encrypt and exchange intermediate results, which are used to help calculate gradients and loss values.
- Regional pandemic management centers and individual airlines calculate the encryption gradient and add additional masks, respectively. Regional pandemic management centers also calculate the encryption loss. The two parties then write the encrypted results into the blockchain's ledger through the SDK.
- Smart contracts on the blockchain decrypt the gradients and losses newly written into the ledger and send the results back to both parties. Regional pandemic management

centers and individual airlines unmask the gradient information and update the model parameters according to the gradient information.

- Meanwhile, the parameters trained by the model are written into the blockchain. Then, the parameters can be read for better pandemic prediction and prevention and can help control policy formulation.

Through the above process, the spread of infection caused by the flight flow between different regions can be calculated. In this study, we used the SIR model (susceptible–infected–removed) as the model to be trained.

3.1. Pandemic Transmission Model Based on SIR

The SIR model mainly simulates the evolution of susceptible population S, infected population I, and removed population R (including those with immunity, those who are no longer infected, or those who have died after being cured). The key parameters of the model are shown in Table 1, where effective reproduction number R is the core parameter trained by federated learning.

Table 1. The specifications of SIR model's symbols

Symbol	Meaning
R	Effective reproduction number (=$\frac{\beta}{\gamma}(1-\frac{C}{N})$); the average number of infections per patient
R_0	Basic reproduction number (=$\frac{\beta}{\gamma}$)
β	Average frequency of exposure (per day)
γ	Average removal frequency (per day)
N	Total population (also used as initial susceptible population)
R_{all}	Final outbreak size (replaced by final recovery)
S_{last}	The rest of the uninfected population
$RMSE$	Root-mean-square error

The SIR model is calculated as follows:

$$\frac{dS}{dt} = -\frac{\beta}{N}IS \quad (1)$$

$$\frac{dS}{dt} = \frac{\beta}{N}IS - \gamma I \quad (2)$$

$$\frac{dR}{dt} = \gamma I \quad (3)$$

where t is time, $S(t)$ is the number of susceptible persons at t, $I(t)$ is the number of infected persons at t and $R(t)$ is the number of cured persons at t.

$$N = S + I + R \quad (4)$$

In the initial state, $S(0) = S_0$, $I(0) = I_0$ and $R(0) = R_0$. According to Formulas (1) and (3):

$$S = S_0 e^{[\frac{\beta}{N\gamma}(R-R_0)]} \quad (5)$$

Eventually, I(t) approaches zero according to Formulas (4) and (5); thus, it can be solved with:

$$R_{all} = N - S_0 e^{[\frac{\beta}{N\gamma}(R_{all}-R_0)]} \quad (6)$$

We accumulate the time series $C(t)$ that is formed by the daily data, as shown in Formula (7).

$$C(t) = I(t) + R(t) = N - S_0 e^{[\frac{\beta}{N\gamma}(R-R_0)]} \tag{7}$$

Therefore, the least squares method is used to conduct regression analysis on the time series formed by the confirmed data and obtain the estimation of each parameter, as shown in Formula (8).

$$\underset{\beta,\gamma}{argmin}(\| C_t - \hat{C}_t(\beta,\gamma,S_0) \|) \tag{8}$$

To protect the patients' private information in the electronic medical record data, we adopted the federal learning method to fine-tune the model training. Formula (8) is the objective function of federated learning.

3.2. Aviation Network Transmission Model

To determine the influence of air traffic network propagation, the probability of cities being affected by diffusion is calculated according to the parameters transmitted by federated learning. The susceptibility probability of flights taking off from each city is approximated as follows:

$$P_{plane} = R_{plane} \cdot \frac{I_{plane}}{N_{plane}} \tag{9}$$

R_{plane} is the effective reproduction number of the flight, I_{plane} represents the presence of cases on the flight and N_{plane} is the total number of people on the flight. Thus, the probability of city k being affected by the imported spread of the pandemic can be calculated, as shown in Formula (10).

$$P_k = 1 - P_k = 1 - \prod_{n=1}^{num_k}[\prod_{m=1}^{f_n}(1 - P_m)] \tag{10}$$

num_k represents the number of cities which have flights to city k, f_n represents the number of flights from city n and P_m is the susceptibility probability of flight m.

Based on the probability of city k being affected, the importance of airports is ranked by measuring the location of airport nodes and the airline flow of airports. A weighted proximity algorithm is used to highlight the influence of path distance between nodes on recognition results.

$$P_i^C = P_k \cdot (M - 1/\sum_{j=1}^{M} f(d_{ij})) \tag{11}$$

M represents the number of airports in the network, and d_{ij} represents the minimum number of transit times from airport i to airport j, where $f()$ is a logarithmic function. The importance output matrix of adjacent nodes is set for evaluation:

$$H = \begin{bmatrix} 1 & \delta_{12}D_2/k^2 & \dots & \delta_{1M}D_M/k^2 \\ \delta_{12}D_2/k^2 & 1 & \dots & \delta_{2M}D_M/k^2 \\ \vdots & \vdots & \dots & \vdots \\ \delta_{M1}D_1/k^2 & \delta_{M2}D_2/k^2 & \dots & 1 \end{bmatrix} \tag{12}$$

In the matrix, δ_{ij} is the contribution allocation parameter. If two nodes are connected, the value is 1; otherwise, the value is 0. The element on the diagonal is 1, which means that the contribution ratio of the node to itself is 1. D_i is the degree of the airport i and k is the average degree of the node:

$$k = \sum_{i=1}^{M} D_i/M \tag{13}$$

Then, the weight S_i of each airport node is calculated:

$$S_i = \sum_{j \in N_i} W_{ij} \quad (14)$$

N_i is the neighbor node set of node i and W_{ij} is the weight of the edge directly connected to node i. According to Formulas (11)–(14), the importance of each airport node is obtained as C_i.

$$C_i = \sum_{i=1, j \neq i}^{M} P_j^C \delta_{ij} S_j D_j / k^2 \quad (15)$$

3.3. *Federated Learning Training Process Assisted by Blockchain*

The workflow of PPChain mainly includes four stages: initializing the collaborative training alliance, writing the encrypted results to the ledger, reading and sending the results, and updating the model parameters.

In the stage of initializing the collaborative training alliance, let us assume that the participants of N_1 (CDC (Centers For Disease Control And Prevention) $p_i, i \in \{1, 2, \ldots, N_1\}$), and the participants of N_2 (airlines $q_i, i \in \{1, 2, \ldots, N_2\}$) join the blockchain network and obtain a configuration file containing predefined information such as a collaboration model and initialization parameters to form a collaborative training alliance. The blockchain network randomly selects $M(M = M_1 + M_2$, where $M_1 = N_1/2, M_2 = N_2/2)$ participants to form a certification consortium to enable the system's consensus algorithm.

It is a remarkable fact that smart contracts will be deployed on the blockchain to create a key pair for each entity that joins the network and will return the public key as a means of identification and encryption, while the private key will be stored in the blockchain ledger along with the entity's characteristic information. It will be read by the smart contract for decryption during the result reading and sending phase.

To ensure communication efficiency and make collaborative training and node identity information easy to maintain, the channel mechanism is applied to the blockchain network. The channel is a dedicated subnet for specific members to communicate with each other. Different types of transactions will be executed in different task channels. Therefore, a relatively independent ledger will be formed for easy management and maintenance, and finally, communication efficiency will be improved.

In this paper, two types of channels were designed in which the identity channel is used for the storage of private keys and entity characteristic information. We designed an identity form for the registration of information for nodes joining the network. Forms consist of normalized data containing entity identity information that circulates in channels in the form of smart contracts. The typical data structure is shown in Table 2.

Table 2. The typical data structure of an identity form.

Symbol	Explanation
ID	The identity number of the entity
type	Categories of entities (medical, aviation)
name	Name of entity
time	The time that the entity joins the blockchain network
state	Whether the entity is incorporated into an authentication consortium
s_k	The entity's private key information

The other channel is used for the collaborative training of the federated learning method, where relevant entities join and complete business on the channel after authentication and authorization. Additionally, the channel ledger records the complete process of collaborative training. When the training is completed, the ledger data of the channel are hashed to form a hash value and stored in the system's general ledger. Similarly, collaborative training is also implemented via forms in the channel through smart contracts. Table 3 shows the form structure of the co-training process.

Table 3. The form structure of the co-training process.

Symbol	Explanation
$MissionID$	Serial number of the cooperative training task
$Owner$	The entity that provides the gradient
$epoch$	The number of rounds iterated by cooperative training
$param$	Encrypted gradient information
$timestamp$	The timestamp of the gradient written to the ledger
$collaborators$	A collection of entities that participate in the corresponding collaboration
$mask$	Mask information

In the stage of writing the encryption results to the ledger, the entity that uploads the model fills in the corresponding form information and iteratively updates the epoch field. The form is broadcast on the channel. The certification consortium verifies and signs off on transactions. The transactions are handed over to channel members for consensus. If a consensus is reached, the transaction-committing channel is packaged into transaction blocks.

After the above steps, the smart contracts on the blockchain decrypt the gradients and losses newly written to the ledger and send the results back to both parties, which are the results of the reading and sending phase. The pandemic control centers and airlines will uncover the gradient information. In the model parameter update stage, the model parameters are updated according to the gradient information, and the above process can be summarized as the collaborative training step in Table 4.

Table 4. The proceedings of collaborative training.

Step	CDC	Airport	Smart Contracts on Blockchain
1	Initialize β, γ	Initialize R_{plane}	Create an encryption key pair and send the public key to both parties
2	Calculate $\parallel \Gamma_A \parallel = \parallel C_t - \hat{C}_t(\beta, \gamma, S_0) \parallel$ and send it to the blockchain	Calculate $\parallel \Gamma_B \parallel = \parallel C_t - \hat{C}_t(R_{plane}, S_0) \parallel$ and send it to the blockchain	Create the collaboration form and write $\parallel \Gamma \parallel = \parallel \Gamma_A \parallel + \parallel \Gamma_B \parallel$
3	Calculate $\frac{\partial \Gamma}{\partial R_0}$, then encrypt it and write it to the blockchain	Calculate $\frac{\partial \Gamma}{\partial R_{plane}}$, then encrypt it and write it to the blockchain	Read the contents of the ledger, decrypt them and send to both parties
4	Update β, γ	Update R_{plane}	Update block
$Obtainedcontent$	β, γ	R_{plane}	The ledger of co-training results

4. Experiment and Analysis

This section introduces the experimental environment of PPChain and evaluates the system performance from four aspects: training accuracy, time cost, transaction processing efficiency, and transaction latency.

Hyperledger Fabric was used to build the blockchain network in the system, and the federated learning parameters were transmitted through smart contracts. The main body of the system model was realized under the chain, and SDK was used to read and write the data on the chain.

4.1. Experiment Settings

To evaluate the system performance of the proposed PPChain, experiments were run on real datasets. Global pandemic data published by John Hopkins University were used as the pandemic data, and public datasets of flight connections were used as the inter-regional flight information.

In this experiment, the two types of datasets were randomly divided into 10 subsets with equal numbers and assigned 20 participants to create local datasets.

The experiment was run on a laptop equipped with a 4-core, 8-thread Intel CPU i7-12800HX and 32 GB of memory. The Python programming language was used to develop SDKs to implement the business logic of the system. Smart contracts were written in the Go language. The learning model was written using Python 3.9 and PyTorch 1.4.0 and executed on the NVIDIA GeForce GTX 3090Ti GPU.

4.2. Performance Evaluation

Table 5 describes the parameters for co-training.

Table 5. The parameters for co-training.

Parameter	Value
$numberofepochs$	10
$numberofiterations$	1500
$learningrate$	0.015
$batchsize$	64

After collaborative training, the training-obtained R_0 is fitted with the actual infection curve, and the accuracy of the training is judged by $RSME$, as shown in Figure 2. It can be seen that the system can better implement collaborative training.

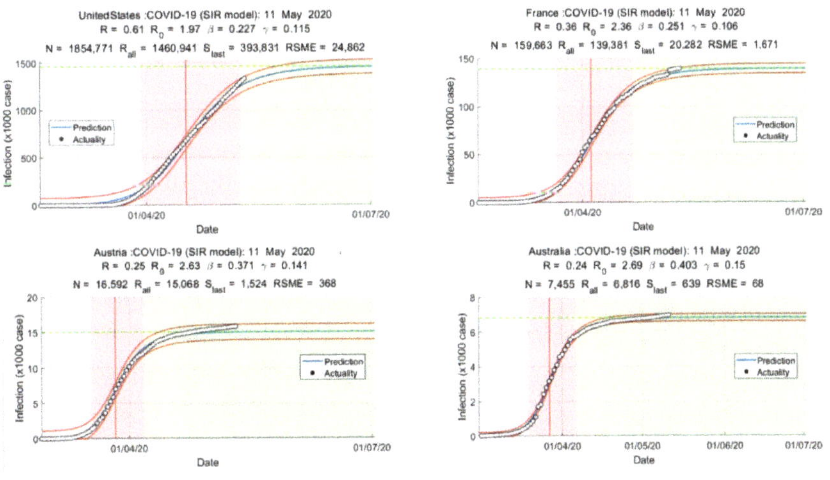

Figure 2. Infection curve fitting graph.

In this paper, a more accurate impact of the aviation network on the spread of the pandemic is obtained through collaborative training because this method realizes the data alignment between common entities without exposing privacy. Based on the R_{plane}, we applied it to the designed aviation network propagation model to obtain a directed network diagram of the pandemic's propagation through the aviation network, as shown in Figure 3.

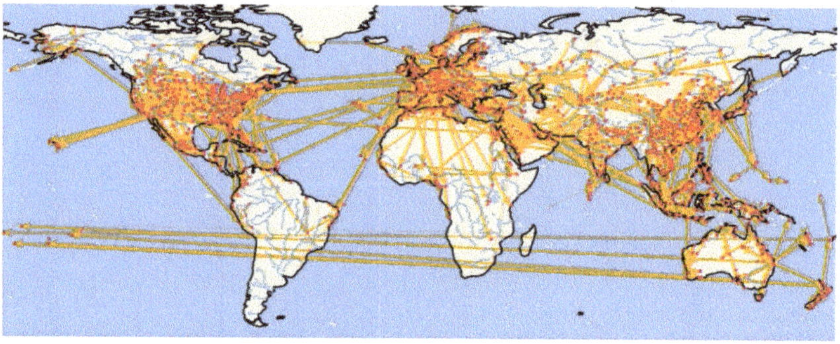

Figure 3. Directed network of pandemic transmission in aviation network.

This paper explores the impact of the aviation network on the COVID-19 pandemic to assist in the formulation of effective prevention and control policies. In this paper, the top 20 airports with the highest importance and the top 20 airports with median importance were selected for comparison.

After airport closures, the data can be applied back into the model that has been trained for calculation. By closing these airports, the number of cities affected daily by the imported spread of the pandemic is as shown in Figure 4. From the results, it can be seen that the effect of closing 20 generally important airports is small, but closing the top 20 most important airports will significantly reduce the number of affected cities.

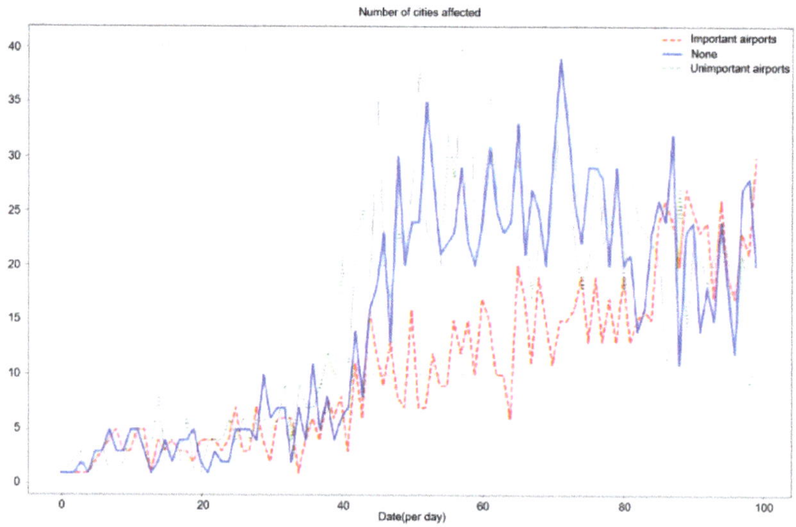

Figure 4. Comparison of the impact of airports on the spread of COVID-19.

Next, the changes in the transaction efficiency and transaction latency in the collaborative training process of the system were recorded. The transaction processing efficiency is reflected by the system running time, which includes the process of reading data, verifying signatures, completing transactions, achieving consensus, and obtaining data on-chain. Transaction latency mainly includes consensus delay and communication delay between nodes and is divided into maximum delay, average delay, and minimum delay due to network jitter.

In this paper, the number of network nodes was changed to test the changes in transaction efficiency and transaction latency, as shown in Figures 5 and 6.

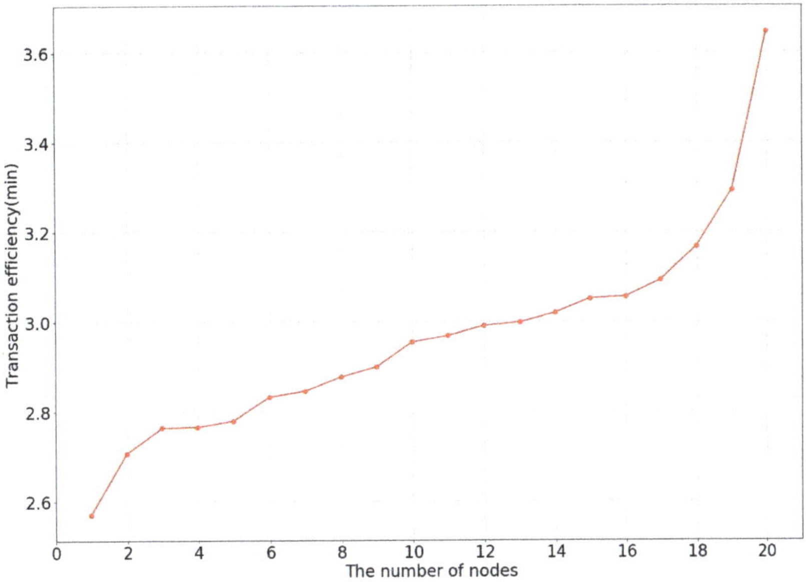

Figure 5. The effect of the number of nodes on transaction efficiency in cooperative training.

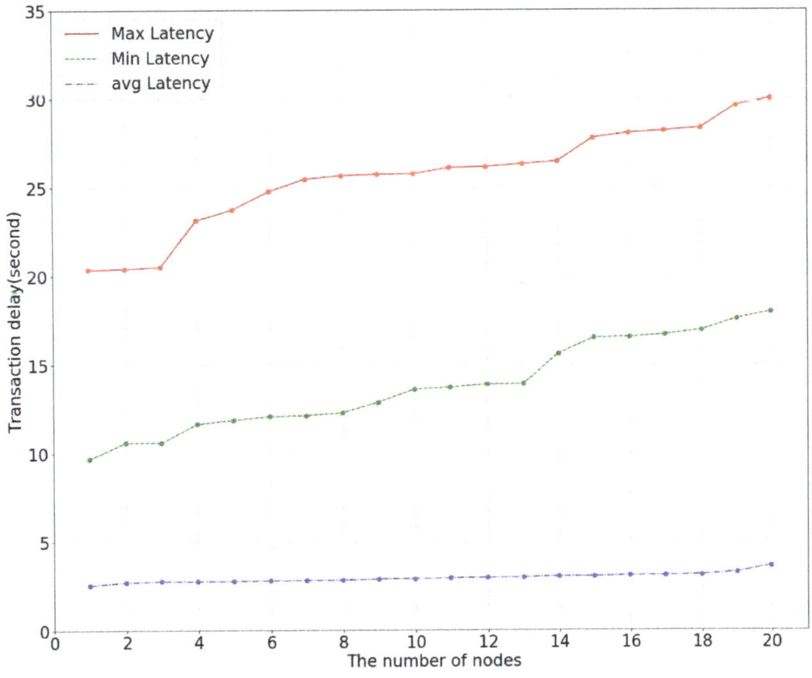

Figure 6. The effect of the number of nodes on transaction delay in cooperative training.

5. Conclusions

In this study, based on the problem that medical centers and airlines cannot fully obtain the flight information of confirmed patients or the illness status of passengers due to privacy protection, we designed PPChain without sharing the original data. The system combines federated learning with a blockchain, improving the security of shared processes without the need for centralized trust. Through experiments, we have verified that effective data-sharing can be achieved without destroying privacy by simulating the impact of aviation policies. Thus, we can formulate prevention and control policies more scientifically and rationally, and predict the spread of a pandemic more accurately.

Author Contributions: Conceptualization, T.C.; methodology, T.C. and Y.P.; software, T.C.; validation, T.C.; formal analysis, T.C.; resources, T.C. and J.Z.; data curation, T.C.; writing—original draft preparation, T.C. and Y.P.; writing—review and editing, T.C.; visualization, Y.P. and T.H.; supervision, H.C. All authors have read and agreed to the published version of the manuscript.

Funding: This research received no external funding.

Informed Consent Statement: Informed consent was obtained from all subjects involved in the study.

Conflicts of Interest: The authors declare no conflict of interest.

References

1. Bavel, V.; Cichocka, A.; Capraro, V. National identity predicts public health support during a global pandemic. *Nat. Commun.* **2022**, *13*, 517. [CrossRef] [PubMed]
2. Raffetti, E.; Mondino, E.; Baldassarre, G. Epidemic risk perceptions in Italy and Sweden driven by authority responses to COVID-19. *Sci. Rep.* **2022**, *12*, 9291. [CrossRef] [PubMed]
3. Brusselaers, N.; Steadson, D.; Bjorklund, K. Evaluation of science advice during the COVID-19 pandemic in Sweden. *Humanit. Soc. Sci. Commun.* **2022**, *9*, 91.
4. Mishra, S.; Scott, J.A.; Laydon, D.J. Comparing the responses of the UK, Sweden and Denmark to COVID-19 using counterfactual modelling. *Sci. Rep.* **2021**, *11*, 16342. [CrossRef]
5. Yang, Q.; Liu, Y.; Cheng, Y.; Kang, Y.; Chen, T.; Yu, H. Vertical Federated Learning. In *Federated Learning*; Synthesis Lectures on Artificial Intelligence and Machine Learning; Springer: Cham, Switzerland, 2020; pp. 69–81.
6. Zhou, X.; Xu, M.; Wu, Y.; Zheng, N. Deep Model Poisoning Attack on Federated Learning. *Future Internet* **2021**, *13*, 73.
7. Qian, F.; Zhang, A. The value of federated learning during and post-COVID-19. *Int. J. Qual. Health Care* **2021**, *33*, mzab010. [CrossRef]
8. Kallel, A.; Rekik, M.; Khemakhem, M. Hybrid-based framework for COVID-19 prediction via federated machine learning models. *J. Supercomput.* **2022**, *78*, 7078–7105. [CrossRef]
9. Salam, M.; Taha, S.; Ramadan, M. COVID-19 detection using federated machine learning. *PLoS ONE* **2021**, *16*, e0252573.
10. Wang, R.; Xu, J.; Ma, Y.; Talha, M.; Al-Rakhami, M.S.; Ghoneim, A. Auxiliary Diagnosis of COVID-19 Based on 5G-Enabled Federated Learning. *IEEE Netw.* **2021**, *35*, 14–20. [CrossRef]
11. Zhang, W.; Zhou, T.; Lu, Q.; Wang, X.; Zhu, C.; Sun, H.; Wang, Z.; Lo, S.K.; Wang, F.-Y. Dynamic-Fusion-Based Federated Learning for COVID-19 Detection. *IEEE Internet Things J.* **2021**, *8*, 15884–15891. [CrossRef]
12. Elshabrawy, K.M.; Alfares, M.M.; Salem, M.A.M. Ensemble Federated Learning for Non-II D COVID-19 Detection. In Proceedings of the 2022 5th International Conference on Computing and Informatics (ICCI), Cairo, Egypt, 9–10 March 2022; pp. 57–63.
13. Alam, U.; Mahbub, R. Federated Semi-Supervised Multi-Task Learning to Detect COVID-19 and Lungs Segmentation Marking Using Chest Radiography Images and Raspberry Pi Devices: An Internet of Medical Things Application. *Sensors* **2021**, *21*, 5025. [CrossRef]
14. Chen, J.J.; Chen, R.; Zhang, X.; Pan, M. A Privacy Preserving Federated Learning Framework for COVID-19 Vulnerability Map Construction. In Proceedings of the ICC 2021—IEEE International Conference on Communications, Xiamen, China, 14–23 June 2021; pp. 1–6.
15. Samuel, O.; Omojo, A.B.; Onuja, A.M.; Sunday, Y.; Tiwari, P.; Gupta, D.; Hafeez, G.; Yahaya, A.S.; Fatoba, O.J.; Shamshirband, S. IoMT: A COVID-19 Healthcare System driven by Federated Learning and Blockchain. *IEEE J. Biomed. Health Inform.* **2023**, *8*, 823–834. [CrossRef]
16. Pang, J.; Huang, Y.; Xie, Z.; Li, J.; Cai, Z. Collaborative City Digital Twin for the COVID-19 Pandemic: A Federated Learning Solution. *Tsinghua Sci. Technol.* **2021**, *26*, 759–771.
17. Nakamoto, S. Bitcoin: A peer-to-peer electronic cash system. In Proceedings of the Nakamoto 2008 BitcoinAP, Nakamoto, Japan, 31 October 2008.
18. Maesa, D.F.; Damiano; Mori, P. Blockchain 3.0 applications survey. *J. Parallel Distrib. Comput.* **2020**, *138*, 99–114. [CrossRef]
19. Dai, H.; Zheng, Z.; Zhang, Y. Blockchain for Internet of Things: A Survey. *IEEE Internet Things J.* **2019**, *6*, 8076–8094. [CrossRef]
20. Wu, C.; Xiong, J.; Xiong, H.; Zhao, Y.; Yi, W. A Review on Recent Progress of Smart Contract in Blockchain. *IEEE Access* **2022**, *10*, 50839–50863. [CrossRef]

21. Chen, J.; Xia, X.; Lo, D.; Grundy, J.; Luo, D.X.; Chen, T. Defining Smart Contract Defects on Ethereum. *IEEE Trans. Softw. Eng.* **2022**, *48*, 327–345. [CrossRef]
22. Yamashita, K.; Nomura, Y.; Zhou, E.; Pi, B.; Jun, S. Potential Risks of Hyperledger Fabric Smart Contracts. In Proceedings of the 2019 IEEE International Workshop on Blockchain Oriented Software Engineering (IWBOSE), Hangzhou, China, 14 March 2019; pp. 1–10.
23. Mallikarjuna, B.; Shrivastava, G.; Sharma, M. Blockchain technology: A DNN token-based approach in healthcare and COVID-19 to generate extracted data. *Expert Syst.* **2022**, *39*, e12778. [CrossRef] [PubMed]
24. Jiang, C.J.; Wu, H.; Yang, Y.; Ma, M.; He, J. BlocHIE: A BLOCkchain-Based Platform for Healthcare Information Exchange. In Proceedings of the 2018 IEEE International Conference on Smart Computing (SMARTCOMP), Taormina, Italy, 18–20 June 2018; pp. 49–56. [CrossRef]
25. Jiang, S.; Cao, J.; Wu, H.; Chen, K.; Liu, X. Privacy-preserving and efficient data sharing for blockchain-based intelligent transportation systems. *Inf. Sci.* **2023**, *635*, 72–85. [CrossRef]
26. Chen, X.; Ji, J.; Luo, C. When machine learning meets blockchain: A decentralized, privacy preserving and secure design. In Proceedings of the IEEE International Conference on Big Data, Pisca Taway, NJ, USA, 10–13 December 2018; pp. 1178–1187.
27. Ramanan, P.; Nakayama, K. BAFFLE: Blockchain based aggregator free federated learning. In Proceedings of the 2020 IEEE International Conference on Blockchain (Blockchain), Rhodes, Greece, 2–6 November 2020; pp. 72–81.
28. Rehman, M.H.; Salah, K.; Damiani, E. Towards blockchain based reputation-aware federated learning. In Proceedings of the International Symposium on Edge Computing Security and Blockchain, Toronto, ON, Canada, 6 July 2020; pp. 183–188.
29. Islam; Amin, A.A.; Shin, S.Y. FBI: A Federated Learning-Based Blockchain-Embedded Data Accumulation Scheme Using Drones for Internet of Things. *IEEE Wirel. Commun. Lett.* **2022**, *11*, 972–976. [CrossRef]
30. Zhu, J.; Cao, J.; Saxena, D.; Jiang, S.; Ferradi, H. Blockchain-empowered Federated Learning: Challenges, Solutions, and Future Directions. *ACM Comput. Surv.* **2023**, *55*, 240. [CrossRef]

bioengineering

MDPI

Article

New Approach for Generating Synthetic Medical Data to Predict Type 2 Diabetes

Zarnigor Tagmatova [1], Akmalbek Abdusalomov [1,*], Rashid Nasimov [2], Nigorakhon Nasimova [2], Ali Hikmet Dogru [3] and Young-Im Cho [1,*]

1 Department of Computer Engineering, Gachon University, Sujeong-Gu, Seongnam-Si 461-701, Republic of Korea
2 Department of Artificial Intelligence, Tashkent State University of Economics, Tashkent 100066, Uzbekistan
3 Department of Computer Science, University of Texas at San Antonio, San Antonio, TX 78249-0667, USA; alihikmet.dogru@utsa.edu
* Correspondence: bobomirzaevich@gmail.com (A.A.); yicho@gachon.ac.kr (Y.-I.C.)

Abstract: The lack of medical databases is currently the main barrier to the development of artificial intelligence-based algorithms in medicine. This issue can be partially resolved by developing a reliable high-quality synthetic database. In this study, an easy and reliable method for developing a synthetic medical database based only on statistical data is proposed. This method changes the primary database developed based on statistical data using a special shuffle algorithm to achieve a satisfactory result and evaluates the resulting dataset using a neural network. Using the proposed method, a database was developed to predict the risk of developing type 2 diabetes 5 years in advance. This dataset consisted of data from 172,290 patients. The prediction accuracy reached 94.45% during neural network training of the dataset.

Keywords: synthetic medical data; type 2 diabetes; prediction of diseases; shuffling

Citation: Tagmatova, Z.; Abdusalomov, A.; Nasimov, R.; Nasimova, N.; Dogru, A.H.; Cho, Y.-I. New Approach for Generating Synthetic Medical Data to Predict Type 2 Diabetes. *Bioengineering* **2023**, *10*, 1031. https://doi.org/10.3390/bioengineering10091031

Academic Editors: Hongqing Yu, Alaa AlZoubi, Yifan Zhao and Hongbo Du

Received: 31 July 2023
Revised: 28 August 2023
Accepted: 30 August 2023
Published: 1 September 2023

1. Introduction

Currently, AI algorithms are widely used in medicine to solve many problems, such as classification, disease prediction, risk evaluation, medical image segmentation, and image detection. However, most AI algorithms, particularly deep learning (DL) methods, require considerable data. The size of the dataset plays a crucial role in increasing the accuracy of the algorithms. The highest-performance milestone algorithm relies on a large database. Moreover, there is also a large open-access database, such as ImageNet, containing more than 1.5 million images, and the Open Image Dataset includes more than 9 million data points. However, in the medical field, although there are millions of data collected from various major hospitals around the world, there are still very few databases open to the public. As medical data include the personal information of patients, they cannot be shared for maintaining privacy. However, certain programmers are allowed to use them, keeping the identity of the patients confidential.

In existing algorithms, the number of recorded patients is either not satisfactory for training DL algorithms or not labeled. For example, the Pima Indian Diabetes Database, which is the most widely used program for training diabetes detection algorithms, contains data from only 768 patients (females). Annual survey: The Behavioral Risk Factor Surveillance System (BRFSS) of the US consists of more than 100,000 annual patient records, but the data are incomplete. Furthermore, finding an appropriate database for training AI models is difficult. In particular, it is almost impossible to find a database required to train predictive algorithms because the data collection process is time consuming and difficult. For example, there is hardly any open-access database that can be used to predict diabetes or its potential complications 5–6 years ahead. Therefore, over the past decade, considerable research has been conducted on the development of synthetic medical data. There are many reviews of these works [1–5].

Typically, there are three types of synthetic data: fully synthetic, semisynthetic, and hybrid. To develop a semi-synthetic database, the main features and statistical distribution of the specific dataset were imitated, that is, a new database that preserves the statistical distribution of the real one was developed. The main purpose of this method is to hide patient data from real databases and thus make the data close to that in the public database open to everyone without compromising privacy. The second method is a hybrid method, in which a large database is developed using a specific small dataset. The goal of such methods is to synthetically increase the amount of data in a small dataset via data augmentation. The third method involves developing a completely new dataset without using a real database. This method is typically used in cases where a database is unavailable in a particular field.

While studying the literature, it is clear that most studies have been conducted in the first and second directions. Only a few methods have been developed in the third research area, most of which are highly complex. In addition, most of these studies were aimed at developing electronic health records, and almost no research has been conducted on predicting the disease in advance.

Furthermore, the traditional approach to generating fully synthetic medical data involved manual input from medical professionals, who had to painstakingly extract the necessary rules from the guidelines and books to create datasets. This process was not only time consuming but also prone to errors and inconsistencies. Additionally, it relied heavily on the availability and cooperation of medical personnel, which further hindered its efficiency. To overcome these challenges, our proposed method utilizes neural network algorithms and shuffling techniques. By leveraging these technologies, we can automate the process of generating synthetic medical data with minimal human intervention. This not only saves time but also ensures accuracy and consistency in the generated datasets.

Moreover, our method allows for scalability, making it suitable for large-scale studies that require extensive amounts of data. It also enables researchers to easily customize the generated datasets according to their specific research requirements. This method also addresses privacy concerns associated with sensitive medical data. Since the database is created using statistical information rather than individual patient records, it ensures anonymity while still providing valuable insights into disease identification.

The main concept of this method is as follows. Many types of statistical medical data are currently available. Most of them are open to use, and much of their data are detailed in special reports or in research papers. In the first stage, a primary dataset was developed using these statistics. Subsequently, a neural network (NN) was trained using this dataset. After every five training epochs, the data were shuffled using a special shuffling operation. The developed dataset was saved after each shuffle. The dataset for which NN showed the highest accuracy was selected. From the calculations, it was estimated that the resulting dataset obtained by this method was satisfactory for practical use. The database created using our method can be a valuable tool in training various AI algorithms for identifying and predicting type 2 diabetes 5 years in advance. Today, such database does not exist or is not open for use, but only statistical data are available.

2. Related Works

Medical data can be in the form of images, numbers, texts, or comments. In addition, they can be single data or time-series data. Therefore, the methods proposed for generating synthetic medical data differ. For example, generational neural diffusion models, variation autoencoders, and a generative adversarial network (GAN) are mainly used to develop synthetic medical images/data [6–8], whereas algorithms such as Bayesian networks [9] and classification and regression trees [10,11] are used to develop numerical (quantitative) and non-numerical (qualitative), and recurrent deep learning models are used to build time-series databases [12]. Because this study aims to develop a numerical database, the issues of generating images and developing non-numerical databases are beyond the scope of this study. Further information can be found in [13].

As we examine methods for generating numerical data, it becomes clear that the majority of the works that have been proposed recently are based on real databases. Their main purpose is to increase the security of the real database, hide the patients' personal data, and, more precisely, change the data to such an extent that the patient's identity cannot be recognized; this resembles high-level encryption. Synthetic databases that properly depict the original data distribution, for instance, would significantly minimize patient privacy concerns and may be freely shared in place of the original patient data. To develop a differentially private synthetic database, the authors of [14] presented deep learning algorithms that can capture the relationships between various variables. Ref. [15] has developed a high-fidelity open generator that generates synthetic data using a probabilistic relational model. This generator met certain privacy requirements and produced an imitation of the large French insured patients (SNDS) database. The most common method for generating synthetic data based on real data is to use GAN. Various modifications of GANs have been used to generate synthetic medical data, including HC_GAN, medWGAN, AC_GAN, MC_medGAN, EMR_WGAN [16], and EEG_GAN [17]. A synthetic replica of numerous databases was produced as a result of the studies described above and made accessible for public use, such as the NIH National COVID Cohort Collaborative (N3C), the CMS Data Entrepreneur's Synthetic Public Use files, and synthetic variants of the French public health system claims and hospital dataset (SNDS) [18–20].

However, little research has been conducted to construct fully synthetic data. Rubin was the first to propose a method to develop a fully synthetic database [21]. Ragunathan et al. (2003) proposed methods based on combining point and variance estimates from multiple synthetic datasets that were closely related but slightly different from the combining rule for multiple nonresponse imputations [22]. Later, Drechsler et al. developed an improved version of this method to overcome its shortcomings [23]. Walonski developed a method for replicating health records using statistical data and medical records [24]. A statistically valid random shuffle method was developed to increase the cardinality of the heart failure dataset [25]. Although this is not a fully synthetic method, it is likely to be an image-augmentation method. Most of the aforementioned methods for developing a full synthetic dataset are complex; therefore, we propose a relatively easy method in this study.

3. Methodology

When we searched for studies on diabetes prediction or risk factor assessment, we found many statistical studies. Most of these authors have conducted large surveys to assess the risk of diabetes 5–10 years in advance to study the factors that lead to diabetes. For example, in [26], 260,000 people; in [27], 63,000 people; and in [28], more than 93,000 people were surveyed/observed over 5 to 10 years. The most important aspect of these studies is the relationship between risk factors and their possibility of causing the disease is written down in the smallest detail.

After studying these papers, we developed a synthetic database based on the statistics presented in this study. The steps for implementing this concept are illustrated graphically in Figure 1.

1. Development of a primary database based on statistical rules and the given statistical data. Based on the selected statistics, the general statistical data were converted into individual patient data. Further details are provided in Section Developing Primary Database.
2. Primary and Secondary shuffling of the data. Usually, the distribution of the data in the initially generated database is highly unbalanced; therefore, it is difficult to bring them to the desired point using the proposed shuffling algorithm during the training process. A primary shuffle should be performed to distribute data uniformly in the database; this method is discussed in detail in Section 4.
3. The database was fed into a loop consisting of the main shuffling and training processes; this step is the main part of our proposed method, in which the primary and

secondary shuffled dataset is trained on the neural network. The dataset was shuffled using a special shuffle function and fed again to the neural network depending on the test accuracy value of the neural network; this process was continued until a satisfactory value was obtained for the neural network. The database with the highest performance was selected and saved every time it was tested in the neural network.

4. Evaluation of trained data, and performance enhancement. We will discuss these processes in detail in Section 5.

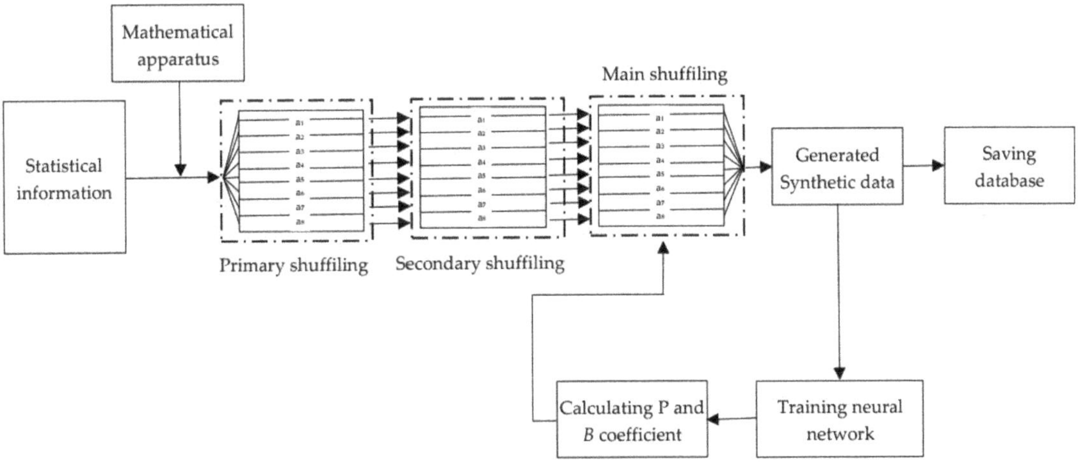

Figure 1. Block diagram of shuffling algorithm.

Developing Primary Database

We have chosen the case in [28] for this study; this work contains the most detailed information and interconnections of the data. In this paper, the China Cardio Metabolic Disease and Cancer Cohort Study survey was analyzed. This survey was conducted with 93,781 non-diabetic participants nationwide between 2011 and 2016. They examined 14 risk factors for diabetes. These risk factors include education, occupation, unhealthy diet, physical inactivity, current alcohol consumption, current smoking, poor sleep, general or central obesity, insulin resistance, prediabetes, hypertension, and dyslipidemia, gender, age. Age dependence of 13 risk factors was studied in groups, namely (40–55, 55–65, 65–75, and $\geq$75 years old).

However, we selected 8 factors for this study: age, gender, unhealthy diet, physical inactivity, current alcohol consumption, poor sleep, general or central obesity, and hypertension. We have carefully selected eight factors for our study because they are easily accessible and can be compiled into a comprehensive database. This allows for future verification and comparison, ensuring transparency and credibility. To be more precise, the factors we chose in this work were chosen not because they were the most important, but because they were easy to implement as a proof of concept. That is, the approach does not underestimate the significance of other potentially critical factors, such as insulin resistance and prediabetes. However, just it takes into account that, it is not always possible to find such data of patients, especially data within 5 years. Therefore, if we used all the factors in the article, the possibility of comparing our work with the real database and estimating accuracy would be reduced. Thus we used easily available risk factors. Once a proof of concept has been established using easily available factors, we can expand our database by incorporating more elements in subsequent phases effortlessly, as our method is easy to implement. Moreover, if we want, we can generate a database with more than 14 risk factors, for example, we can add factors such as ethnicity and cardiac disorder. However, for this we should use other survey with more statistical information.

First, the number of patients in each age group was determined. The average age of the patients, standard deviation, and age limits are given for each group. Patient age was calculated using the following formula:

$$F(x) = \frac{1}{\sqrt{2\pi}} \int_{x_1}^{x_2} e^{-t^2/2} dt \quad (1)$$

where x_1 and x_2 are the age boundaries for each group at t-time in the year. We used the following approach to generate the remaining data. The hazard ratios were calculated for each parameter. For incident diabetes associated with risk factors and risk scores according to age group, hazard ratios (HRs) and 95% confidence intervals (Cis) were calculated using Cox proportional hazards models. It is known that the hazard ratio is determined by the following formula:

$$HR = \frac{\left(\frac{O_d}{E_d}\right)}{\frac{O_h}{E_h}} \quad (2)$$

where O_d is the observed number of events in the group of diabetes, O_h is the observed number of events in the group of healthy people, E_d is the expected number of events in the group of diabetes, and E_h is the expected number of events in the healthy group. Formula (2) was used to arrange the statistical values into groups. It is important to say that, from the beginning, we divided the patients' data in each age group into 2 subgroups: those who developed diabetes within 5 years and those who did not develop diabetes within 5 years/healthy patients. This study aimed to prevent changes in the statistical values of the dataset. Next, we recorded the values in each subgroup based on Formula (2), using the ratio given in the statistics for diabetic and healthy people.

Although in these statistics, the numbers of men and women were given for each age group, they were adjusted for gender. In other words, the significance of gender has not yet been studied. When we analyzed other studies on the importance of gender, we ascertained that the relationship between diabetes and gender was still under investigation. The risk of developing diabetes among women and men depends on nationality and age [20]. Although it is more common among men in the US, it is also more common among women in East Asian countries. For this reason, we did not use the HR formula to develop the gender data but directly developed the data based on the numbers. As we divided the age group into subgroups and information on the proportion of women and men for subgroups was not given, we also kept the age group proportion within the subgroups.

In addition, it is important to note that all of our data were in the form of zero or one, except for age. Information about the age of the participants was in the range of 40–100. Age was divided by 100 to normalize them to other data.

4. Shuffling the Data

4.1. Primary and Secondary Shuffling

When the synthetic data were generated based on statistics, it was observed that the data were unevenly distributed. If this database is fed into a loop consisting of a neural network and a special shuffle function, it is possible to develop a database with the desired form. Therefore, primary shuffling was performed before feeding into the neural network. This step includes two shuffling methods: primary and secondary. As the primary shuffling method, we used the Fisher–Yates shuffling method; as the secondary shuffling method, we used a new simple shuffling method. Because the dataset consists mainly of zeroes and ones, shuffling using the Fisher–Yates method does not produce the expected result. Therefore, in addition to the Fisher–Yates shuffle method, we used a secondary shuffling method. This secondary shuffle method operates according to Formulas (3) and (4):

$$a[i] = a[semilength - i] \quad (3)$$

$$a[semilength] = a[i] \tag{4}$$

4.2. Main Shuffling Algorithm

The main shuffling process begins after the primary and secondary shuffling processes. It is worth noting that while the primary and secondary shuffling processes are accomplished only once, the main shuffling process is accomplished in every cycle.

The input data of the function are subgroup data, and their main parameters are the percentage (P parameter) and starting point (B parameter). Based on these parameters, the function shuffles the data within a specific interval of the incoming subgroup database. More precisely, the starting point of the part to be shuffled was determined by parameter B, and its ending point was determined by parameter P. Parameter P determines the percentage of the length of the data to be shuffled. Each time a database is trained and evaluated in a neural network, its value is modified by a special function depending on the accuracy of the network. We initially set these two parameters to 0 and 0.1. This special function is similar to that of the Adam optimizer with slight modifications and is defined by the following formula:

$$v_t = \beta_1 * v_{t-1} - (1 - \beta_1) * g_t \tag{5}$$

$$s_t = \beta_2 * s_{t-1} - (1 - \beta_2) * g_t^2 \tag{6}$$

$$\Delta\omega_t = \left| \eta \frac{v_t}{\sqrt{s_t + \epsilon}} * g_t \right| \tag{7}$$

$$\omega_{t+1} = \omega_t + \Delta\omega_t \tag{8}$$

Here,

η : **Initial learning rat**
g_t : **Gradient at time t along ω^j**
v_t : **Exponential average of gradients along ω_j**
s_t : **Exponential average of squares of gradients along ω_j**
$\beta_1\beta_2$: **Hyperparameters**

We gave the following values to the hyperparameters: Initial Learning Rate = 1, $\beta_1 = 0.95$, $\beta_2 = 0.99$, $\epsilon = \mathbf{0.0001}$. The coefficient $\boldsymbol{B}$ is increased by 0.12 after each cycle. If the value of coefficient $\boldsymbol{B}$ exceeds 1, the operation $\boldsymbol{B} = \boldsymbol{B} - \mathbf{1}$ will be performed. If the endpoint exceeds the length of the array, the value of coefficient $\boldsymbol{B}$ is set to zero.

After the start and end points of the shuffling were determined, the numbers in between were shuffled using the main shuffle function. Let the data be expressed in the form of matrix $\boldsymbol{A}$ where $\boldsymbol{A} = \begin{vmatrix} a_{11} & a_{21} & \cdots & a_{81} \\ a_{12} & a_{22} & \cdots & a_{81} \\ \cdots & \cdots & \cdots & \cdots \\ a_{1i} & a_{2i} & \cdots & a_{8i} \end{vmatrix}$. All elements of the array, except for the first, were 0 or 1. The last row shows the diagnostic values; therefore, the main shuffle function is only performed on columns 2–7 for each column separately. The values in the first column were shuffled according to the French–Yates algorithm in a given interval. The main shuffling function is defined as follows:

$$a_{out}(j,k) = \begin{vmatrix} \mathbf{1} \\ \mathbf{1} \\ \cdots \\ \mathbf{1} \end{vmatrix} - \begin{vmatrix} a_{j1} \\ a_{j1} \\ \cdots \\ a_{jk} \end{vmatrix} \tag{9}$$

Here, $a_{j,k}$—is the jth column of the subgroup data obtained in the given interval. This function converts 0 s to 1 s and 1 s to 0 sin in the given interval. However, the statistical distribution of the database was distorted. To avoid this, the number of 1 s in a given interval is compared with the previous interval as follows:

$$\Delta a(j,m) = \sum_{m=0}^{k} a_0(j,m) - \sum_{m=0}^{k} a_{out}(j,m) \tag{10}$$

$$N = |\Delta a(j,m)|$$

If $\Delta a(j,m)$ is greater than 0, this means there are N more 0 s converted to 1 s than 1 s converted to 0 s. Therefore, the next interval is selected from the endpoint to the end of the array, and the N interval within this interval is converted to zero. If $\Delta a(j,m)$ is less than zero, the reverse operation will be carried out, that is, N zeroes within this interval are converted to one. Therefore, the statistical distribution of the database was preserved.

4.3. *Training for Loop*

The database contained information on two categories of patients: 87,610 patients were followed up and remained healthy; 6171 patients developed diabetes during this time. As shown, the dataset was highly unbalanced. Typically, neural networks are likely to overfit when trained using such databases. Therefore, we multiplied the data for a small number of classes several times. Considering that the data are not real and unique and will be shuffled several times in the next step, we simply added the same data several times. After increasing the data, the number of people with diabetes was 84,692. Subsequently, the data were used for the next step. It is worth mentioning once again that the data that passed through the first step were not in the form of a single dataset but in the form of separate subgroups. The number of data points in these subgroups is presented in Table 1.

Table 1. Information about subgroup data.

Age Group, Years	**40 to <55**		**55 to <65**		**65 to <75**		**≥75**	
Subgroup	Did not develop diabetes	Developed diabetes	Did not develop diabetes	Developed diabetes	Did not develop diabetes	Developed diabetes	Did not develop diabetes	Developed diabetes
Number of patients	42,825	2306	31,355	2478	11,570	1176	1851	203
Number of male patients	12,505	673	10,786	852	4766	484	795	87
Number of female patients	30,320	1633	20,569	1626	6804	692	1056	116

The second step consists of two processes: shuffling the data and training the neural network on the prepared data. A special main function is used for shuffling.

After shuffling, the subgroup data were combined into a single database and divided by a ratio of 8:2 into training and test databases. The training dataset was then transferred to the NN, which comprised 16, 64, and 32 consecutive hidden layers. After each linear layer, a ReLU activation layer was formed. The neural network was trained over five epochs and evaluated using the test dataset. These two processes (shuffling and training) were repeated until the desired results were achieved.

5. Results and Discussion

In each cycle, the neural network was trained for five epochs, and its performance was assessed using a test database. The test accuracy was sent to the optimizer to determine the percentage and starting points. When the new values of the percentage and starting point are determined, the cycle starts again, and the data are reshuffled and transmitted to the

NN. This cycle was repeated 100 times. The training, validation, and test accuracies at the end of each cycle are shown in Figure 2. A database was saved after each cycle. Among these, the one with the highest number of test results was selected. When training NN with this dataset, the training accuracy was 100%, and the test accuracy was 94.4%.

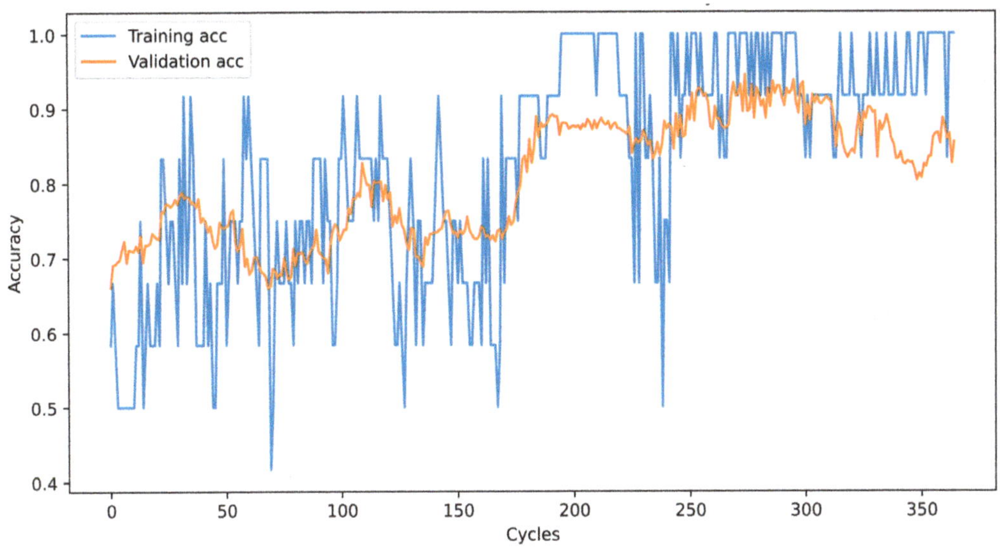

(**a**)

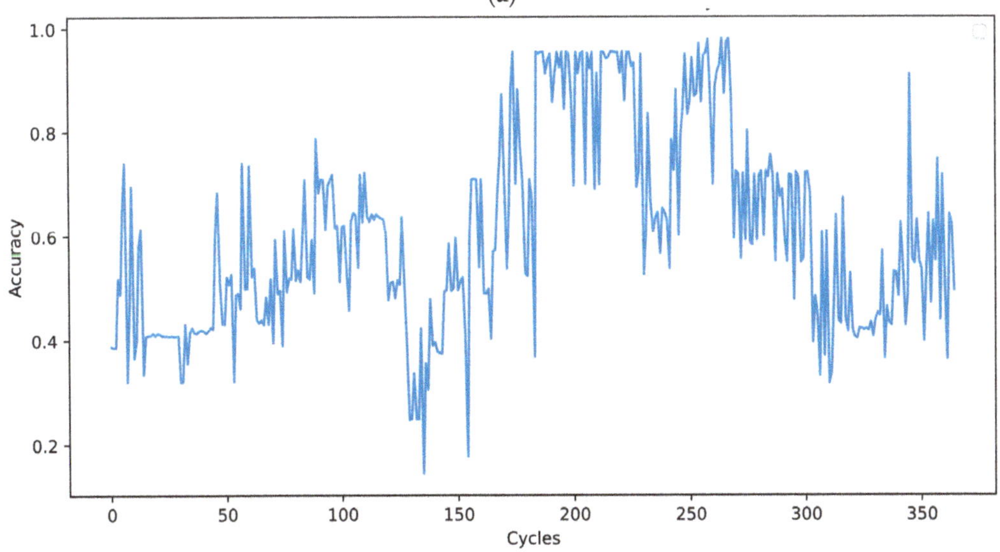

(**b**)

Figure 2. Training, validation (**a**) and test (**b**) accuracy at the end of each training cycle.

As mentioned above, at the end of each cycle, the P- and B-coefficient values change according to (7). Their values over 100 cycles are shown in Figure 3.

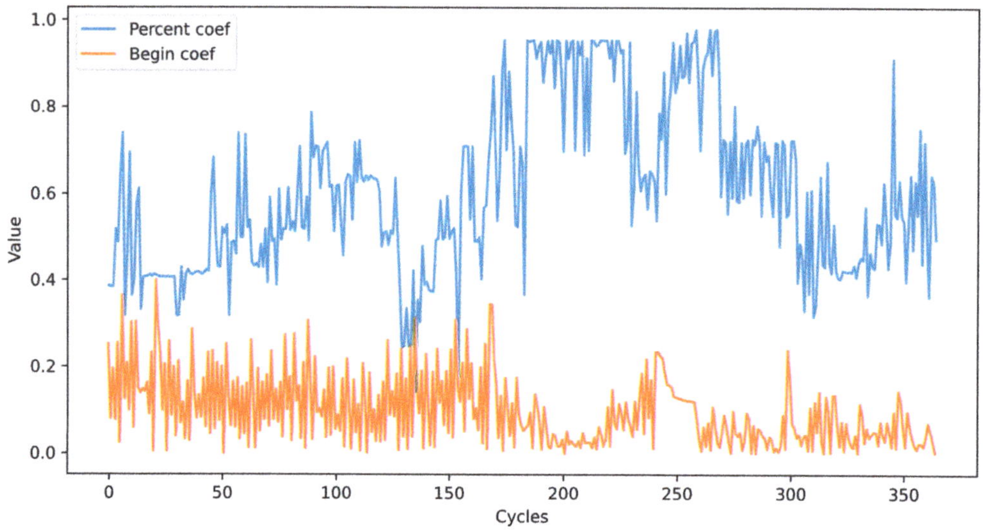

Figure 3. Values of P and B coefficients at the end of each cycle.

5.1. Evaluation of the Method

Various methods have been proposed to evaluate synthetic databases. However, because the proposed method is completely synthetic, these evaluation methods are unsuitable for evaluating our method. The following methods were used to evaluate the method.

First, the database that exhibited the best results when the neural network that was trained was selected. Subsequently, the selected database was trained in a completely new neural network to evaluate the real value of the dataset, as this dataset was intended to train the AI algorithms. The network architecture is presented in Table 2, and the training process is illustrated in Figure 4. The database was split randomly with random state 46, in an 8:2 portion, into the training and test sets.

Table 2. Layers parameters of neural network.

#	Layer Name	Properties
1	Input layer	8 nodes
2	Hidden layer	16 nodes
3	Dropout layer	0.2 coefficient
4	ReLu	
5	Hidden layer	64 nodes
6	Dropout layer	0.2 coefficient
7	ReLu	
8	Hidden layer	32 nodes
9	Dropout layer	0.2 coefficient
10	ReLu	
11	Hidden layer	16 nodes
12	ReLu	
13	Hidden layer	1 nodes
14	Output (Sigmoid) layer	

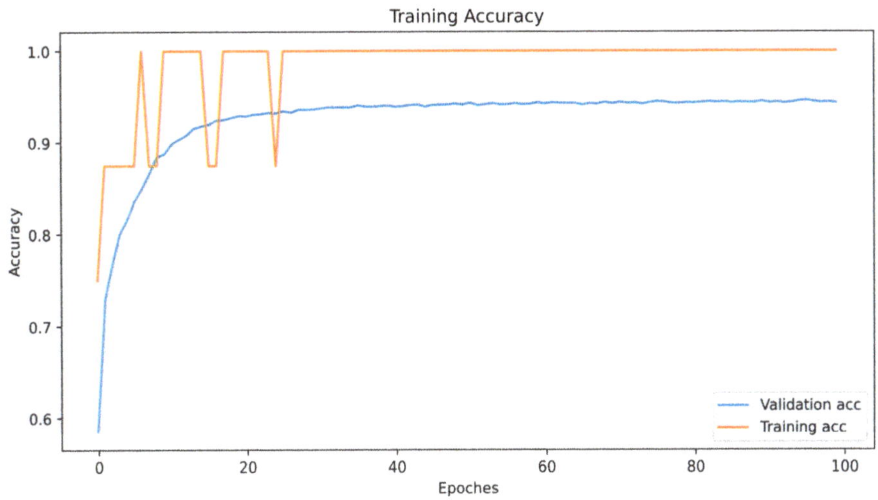

Figure 4. Training and validation accuracy of the neural network trained on the generated dataset.

As shown in Figure 3, the test accuracy reached 90% at the 10th epoch; while at the 20th, it reached 94%; and after that, it remained approximately the same. We extracted the dataset that showed the highest accuracy (94.4%) in these cycles and uploaded it to Kaggle for public use.

Second, they were plotted in the form of a histogram to determine the distribution of the data in the database. For this purpose, we divided the data into two groups: those who developed diabetes within five years and those who remained healthy. We then obtained two matrices of the forms (84,692, 8) and (87,598, 8). The first seven columns of these matrices show the disease risk factors and the last column shows the diagnosis. We initially added seven columns, separated the resulting columns by values, and represented them in the form of 100 histograms (Figure 4). For a comparative study of the data distribution, histograms of the dataset before shuffling and the dataset with the highest results during shuffling are shown in Figure 5.

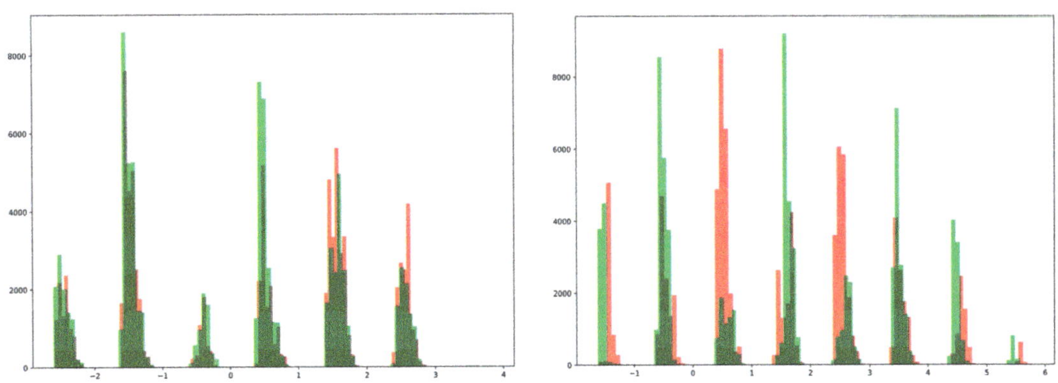

(a) (b)

Figure 5. Data distribution in raw dataset (**a**) and generated dataset (**b**).

As can be seen from the above, the data of different classes in the synthetic dataset generated by the proposed method were well distinguished from each other, and because of this, it was possible to achieve high accuracy when using them to train the neural network.

5.2. Discussion

In this study, we propose a method for generating a synthetic dataset without complex mathematical operations or an initial database. This technique is superior to the previously proposed strategies in numerous respects.

Firstly, most works [22,23] that generated a fully synthetic database calculated the desired statistical distribution from the real database and then used it to generate synthetic data. However, it is important to note that using real datasets in research can be challenging due to privacy concerns and legal restrictions. Accessing and utilizing such datasets may require permissions and agreements that are not always easy to obtain. Therefore, having a method that does not rely on real datasets would greatly simplify the process and make it more accessible for researchers [29–32].

While our proposed method utilizes only the given statistics of the desired database, by using the proposed method with these given statistics, we can generate synthetic data that closely resembles the original dataset without compromising privacy or legal constraints.

Moreover, this approach allows for greater control over the generated data. Researchers can manipulate and experiment with different scenarios by adjusting the statistical parameters provided. This flexibility enables them to explore various possibilities and test hypotheses without being limited by an existing dataset.

Secondly, although the work in [24] has several advantages and is a highly reliable method, it uses a complicated method. In order to create the dataset, medical conclusions and guidelines were used. While this ensures accuracy and credibility, extracting the correct and necessary medical instructions and rules can be both time consuming and expensive. One of the main challenges in using medical conclusions and guidelines is their sheer volume. The vast amount of information available makes it difficult to sift through and extract only what is relevant for creating the dataset. This requires extensive research and analysis, which can be a time-consuming process.

Moreover, obtaining accurate medical instructions and rules often involves consulting experts in the field. These experts may charge high fees for their services, making it expensive to gather the necessary information for creating the dataset. Additionally, medical knowledge is constantly evolving with new research findings and updated guidelines being published regularly. This means that maintaining an up-to-date dataset requires continuous effort and investment.

However, our proposed method offers the possibility of generating databases in a semi-automatic way with minimal human intervention, and most importantly, without the involvement of medical professionals. Thus, it can significantly reduce the time and effort required. Moreover, eliminating the involvement of medical professionals further streamlines the process. Additionally, cost plays a crucial role in any project implementation. Traditional methods involving medical professionals can be expensive due to their expertise and time commitment. However, with our semi-automated approach, costs are significantly reduced as there is no need for specialized personnel or extensive training. Furthermore, speed is a critical factor in today's fast-paced world. Our proposed method ensures the rapid generation of databases. This enables researchers and programmers to access up-to-date databases promptly for analysis and decision-making purposes.

Another advantage is that, in contrast to [25], we used a rule-based shuffle method instead of a random shuffle method to achieve this goal; this helped us achieve our goals faster.

Now, if we turn to the issue of evaluating the quality of the developed method, it is known that today various methods have been developed for the evaluation of synthetic data. However, some are designed to evaluate synthetic images [33], others to determine the level of security [34], and others to evaluate the difference between the distributions of synthetic and real images [35]. However, none of the above methods were suitable for evaluating the proposed method. As we do not have a real database, there is no privacy issue, and simultaneously, there is no possibility of comparing the statistical distribution with that of the real dataset. In such cases, certain authors have suggested the use of specific evaluation methods. For example, a unique evaluation method was used in the work [24], and some statistical data in the database were compared with those of other real statistical information. Similarly, we used a unique approach to evaluate the results of our study. Our goal was to develop a database for training disease classification and prediction algorithms with two main goals. The first was to preserve the statistical distribution of the survey used to construct the dataset, and the second was to ensure that the synthetic data belonging to two different classes were maximally different from each other. In our method, actions at all stages of the proposed method assume the preservation of the statistical distribution of the survey; that is, the generated synthetic dataset is identical to the statistical data of the survey. We expressed the data distribution as a histogram to evaluate the dissimilarity between different classes. As shown in Figure 5, the data in the created database are satisfactorily separated.

One major limitation of the proposed method is relying solely on one survey. Different surveys often focus on different aspects or variables related to a particular topic. Combining these various perspectives allows for a more comprehensive analysis and provides researchers with a broader understanding of the subject matter. Additionally, incorporating data from multiple surveys enhances the generalizability of findings. It helps in identifying patterns and trends across diverse populations or contexts. This broader scope strengthens the validity and reliability of the database. To address this limitation, future research should aim to modify this method to integrate data from multiple surveys seamlessly.

6. Future Work

Currently, much work is being conducted to extract medically important characteristics from existing datasets [36], the main goal of which is to define and evaluate the main risk factors that cause the disease and to use them in disease prediction or diagnosis. By contrast, our proposed method aims to generate a dataset based on given risk factors. In the future, by analyzing the dependence of risk factors and information in the dataset using these two methods, it will be possible to develop an algorithm that determines the relationship between them, which will be an important tool for diagnosis [37–43].

7. Conclusions

In this study, with the help of a special shuffle operator, a synthetic dataset was generated that fully represented the statistical data of the survey conducted by [44] over five years. This database contains two classes: data on patients who developed type 2 diabetes and data on those who remained healthy during a 5-year follow-up. This generated dataset can be used to train AI algorithms designed to predict type 2 diabetes five years in advance. To assess the suitability of the database for this purpose, a neural network was trained using this dataset, and a test accuracy of 94.4% was achieved. From the above, it can be concluded that the accuracy, reliability, and simplicity of the proposed method are important.

While relying on one survey is the limitation of the method, considering information from different surveys is crucial. Future research should focus on creating methods that can encounter several research/survey papers' information to enhance accuracy, comprehensiveness, generalizability, and reliability of the generated database. In conclusion, the proposed easy and semi-automotive method offers a solution by utilizing a neural network and special shuffling function. This approach not only reduces the difficulty associated with generating synthetic medical data but also provides satisfactory results in a more efficient manner.

Author Contributions: Conceptualization, R.N. and Z.T.; formal analysis, R.N.; algorithms: N.N., A.A.; funding acquisition, Y.-I.C.; investigation, A.H.D. and R.N.; methodology, A.A.; project administration, A.A.; software, A.H.D.; supervision, Y.-I.C.; validation, R.N. and N.N.; writing—original draft, A.H.D., Z.T. and R.N.; writing—review and editing, A.A., R.N. and Y.-I.C. All authors have read and agreed to the published version of the manuscript.

Funding: This study was funded by the Korea Agency for Technology and Standards in 2022, project numbers are K_G012002236201, K_G012002073401 and by the Gachon University research fund of 2023 (GCU-(202307790001)).

Institutional Review Board Statement: Not applicable.

Informed Consent Statement: Not applicable.

Data Availability Statement: Not applicable.

Acknowledgments: The authors would like to thank the editor and anonymous referees for their constructive comments toward improving the contents and presentation of this paper.

Conflicts of Interest: The authors declare no conflict of interest.

References

1. Gonzales, A.; Guruswamy, G.; Smith, S.R. Synthetic data in health care: A narrative review. *PLoS Digit. Health* **2023**, *2*, e0000082. [CrossRef] [PubMed]
2. Kokosi, T.; Harron, K. Synthetic data in medical research. *BMJ Med.* **2022**, *1*, e000167. [CrossRef]
3. Turimov Mustapoevich, D.; Muhamediyeva Tulkunovna, D.; Safarova Ulmasovna, L.; Primova, H.; Kim, W. Improved Cattle Disease Diagnosis Based on Fuzzy Logic Algorithms. *Sensors* **2023**, *23*, 2107. [CrossRef]
4. McDuff, D.; Curran, T.; Kadambi, A. Synthetic Data in Healthcare. *arXiv* **2023**, arXiv:2304.03243. [CrossRef]
5. Surendra, H.; Mohan, H. A Review of Synthetic Data Generation Methods for Privacy Preserving Data Publishing. *J. Sci. Technol. Res.* **2017**, *6*, 95–101.
6. Ali, H.; Murad, S.; Shah, Z. Spot the Fake Lungs: Generating Synthetic Medical Images Using Neural Diffusion Models. In *Artificial Intelligence and Cognitive Science*; Longo, L., O'Reilly, R., Eds.; AICS 2022. Communications in Computer and Information Science; Springer: Cham, Switzerland, 2023; Volume 1662. [CrossRef]
7. Jaen-Lorites, J.M.; Perez-Pelegri, M.; Laparra, V.; Lopez-Lereu, M.P.; Monmeneu, J.V.; Maceira, A.M.; Moratal, D. Synthetic Generation of Cardiac MR Images Combining Convolutional Variational Autoencoders and Style Transfer. In Proceedings of the 2022 44th Annual International Conference of the IEEE Engineering in Medicine & Biology Society (EMBC), Glasgow, UK, 11–15 July 2022; Volume 2022, pp. 2084–2087. [CrossRef] [PubMed]
8. Aljohani, A.; Alharbe, N. Generating Synthetic Images for Healthcare with Novel Deep Pix2Pix GAN. *Electronics* **2022**, *11*, 3470. [CrossRef]
9. Kaur, D.; Sobiesk, M.; Patil, S.; Liu, J.; Bhagat, P.; Gupta, A.; Markuzon, N. Application of Bayesian networks to generate synthetic health data. *J. Am. Med. Inform. Assoc.* **2021**, *28*, 801–811. [CrossRef]
10. Reiter, J. Using CART to generate partially synthetic public use microdata. *J. Off. Stat.* **2005**, *21*, 441–462.
11. Umirzakova, S.; Abdusalomov, A.; Whangbo, T.K. Fully Automatic Stroke Symptom Detection Method Based on Facial Features and Moving Hand Differences. In Proceedings of the 2019 International Symposium on Multimedia and Communication Technology (ISMAC), Quezon City, Philippines, 19–21 August 2019; pp. 1–5. [CrossRef]
12. Mosquera, L.; El Emam, K.; Ding, L.; Sharma, V.; Zhang, X.H.; El Kababji, S.; Carvalho, C.; Hamilton, B.; Palfrey, D.; Kong, L.; et al. A method for generating synthetic longitudinal health data. *BMC Med. Res. Methodol.* **2023**, *23*, 67. [CrossRef]
13. Chen, R.J.; Lu, M.Y.; Chen, T.Y.; Williamson, D.F.; Mahmood, F. Synthetic data in machine learning for medicine and healthcare. *Nat. Biomed. Eng.* **2021**, *5*, 493–497. [CrossRef] [PubMed]
14. Guyet, T.; Allard, T.; Bakalara, J.; Dameron, O. An open generator of synthetic administrative healthcare databases. In Proceedings of the IAS 2021—Atelier Intelligence Artificielle et Santé, Bordeaux, France, 28 June 2021; pp. 1–8; ffhal-03326618f.
15. Ghadeer, G.; Jin, L.; Tingting, Z. A review of Generative Adversarial Networks for Electronic Health Records: Applications, evaluation measures and data sources. *arXiv* **2022**, arXiv:2203.07018.
16. Hartmann, K.G.; Schirrmeister, R.T.; Ball, T. EEG-GAN:generative adversarial networks for electroencephalograhic(EEG) brain signals. *arXiv* **2018**, arXiv:1806.01875.
17. Haendel, M.A.; Chute, C.G.; Bennett, T.D.; Eichmann, D.A.; Guinney, J.; Kibbe, W.A.; Payne, P.R.O.; Pfaff, E.R.; Robinson, P.N.; Saltz, J.H.; et al. N3C Consortium. The National COVID Cohort Collaborative (N3C): Rationale, design, infrastructure, and deployment. *J. Am. Med. Inform. Assoc.* **2021**, *28*, 427–443. [CrossRef]
18. CMS. CMS 2008–2010 Data Entrepreneurs' Synthetic Public Use File (DESynPUF). Available online: https://www.cms.gov/Research-Statistics-Data-and-Systems/Downloadable-Public-Use-Files/SynPUFs/DE_Syn_PUF (accessed on 17 July 2022).
19. SNDS Synthetiques. Systeme National des Donnees de Sante. 2021. Available online: https://documentation-snds.health-data-hub.fr/formation_snds/donnees_synthetiques/ (accessed on 20 January 2022).
20. McPherson, A.R.; Bancks, M.P. Assessment for Gender Differences in Trend in Age at Diagnosis of Diabetes among U.S. Adults, 1999–2020. *Diabetes Care* **2023**, *46*, e76–e77. [CrossRef]
21. Rubin, D.B. Discussion: Statistical Disclosure Limitation. *J. Off. Stat.* **1993**, *9*, 461–468.
22. Raghunathan, T.; Reiter, J.; Rubin, D. Multiple imputation for statistical disclosure limitation. *J. Off. Stat.* **2003**, *19*, 1–16.

23. Drechsler, J. Improved Variance Estimation for Fully Synthetic Datasets. 2011. Available online: https://drupal-main-staging.unece.org/fileadmin/DAM/stats/documents/ece/ces/ge.46/2011/18_Drechsler.pdf. (accessed on 28 May 2023).
24. Walonoski, J.; Kramer, M.; Nichols, J.; Quina, A.; Moesel, C.; Hall, D.; Duffett, C.; Dube, K.; Gallagher, T.; McLachlan, S. Synthea: An approach, method, and software mechanism for generating synthetic patients and the synthetic electronic health care record. *J. Am. Med. Inform. Assoc.* **2017**, *25*, 230–238. [CrossRef] [PubMed]
25. Fassina, L.; Faragli, A.; Lo Muzio, F.P.; Kelle, S.; Campana, C.; Pieske, B.; Edelmann, F.; Alogna, A. A Random Shuffle Method to Expand a Narrow Dataset and Overcome the Associated Challenges in a Clinical Study: A Heart Failure Cohort Example. *Front. Cardiovasc. Med.* **2020**, *7*, 599923. [CrossRef] [PubMed]
26. Iyen, B.; Weng, S.; Vinogradova, Y.; Akyea, R.K.; Qureshi, N.; Kai, J. Long-term body mass index changes in overweight and obese adults and the risk of heart failure, cardiovascular disease and mortality: A cohort study of over 260,000 adults in the UK. *BMC Public Health* **2021**, *21*, 576. [CrossRef] [PubMed]
27. Vashist, P.; Senjam, S.S.; Gupta, V.; Manna, S.; Gupta, N.; Shamanna, B.R.; Bhardwaj, A.; Kumar, A.; Gupta, P. Prevalence of diabetic retinopahty in India: Results from the National Survey 2015-19. *Indian J. Ophthalmol.* **2021**, *69*, 3087–3094. [CrossRef] [PubMed]
28. Wang, T.; Zhao, Z.; Wang, G.; Li, Q.; Xu, Y.; Li, M.; Hu, R.; Chen, G.; Su, Q.; Mu, Y.; et al. Age-related disparities in diabetes risk attributable to modifiable risk factor profiles in Chinese adults: A nationwide, population-based, cohort study. *Lancet Healthy Longev.* **2021**, *2*, e618–e628. [CrossRef] [PubMed]
29. Kuldoshbay, A.; Abdusalomov, A.; Mukhiddinov, M.; Baratov, N.; Makhmudov, F.; Cho, Y.I. An improvement for the automatic classification method for ultrasound images used on CNN. *Int. J. Wavelets Multiresolution Inf. Process.* **2022**, *20*, 2150054.
30. Farkhod, A.; Abdusalomov, A.B.; Mukhiddinov, M.; Cho, Y.-I. Development of Real-Time Landmark-Based Emotion Recognition CNN for Masked Faces. *Sensors* **2022**, *22*, 8704. [CrossRef]
31. Mamieva, D.; Abdusalomov, A.B.; Mukhiddinov, M.; Whangbo, T.K. Improved Face Detection Method via Learning Small Faces on Hard Images Based on a Deep Learning Approach. *Sensors* **2023**, *23*, 502. [CrossRef] [PubMed]
32. Jakhongir, N.; Abdusalomov, A.; Whangbo, T.K. 3D Volume Reconstruction from MRI Slices based on VTK. In Proceedings of the 2021 International Conference on Information and Communication Technology Convergence (ICTC), Jeju Island, Republic of Korea, 19–21 October 2021; pp. 689–692.
33. Abdusalomov, A.B.; Nasimov, R.; Nasimova, N.; Muminov, B.; Whangbo, T.K. Evaluating Synthetic Medical Images Using Artificial Intelligence with the GAN Algorithm. *Sensors* **2023**, *23*, 3440. [CrossRef] [PubMed]
34. Yale, A.; Dash, S.; Dutta, R.; Guyon, I.; Pavao, A.; Bennett, K. Generation and Evaluation of Privacy Preserving Synthetic Health Data. *Neurocomputing* **2020**, *416*, 244–255. [CrossRef]
35. Goncalves, A.; Ray, P.; Soper, B.; Stevens, J.; Coyle, L.; Sales, A.P. Generation and evaluation of synthetic patient data. *BMC Med. Res. Methodol.* **2020**, *20*, 108. [CrossRef] [PubMed]
36. Nasimov, R.; Nasimova, N.; Muminov, B. Hybrid Method for Evaluating Feature Importance for Predicting Chronic Heart Diseases. In Proceedings of the 2022 International Conference on Information Science and Communications Technologies (ICISCT), Tashkent, Uzbekistan, 3–5 November 2022; pp. 1–4. [CrossRef]
37. Nodirov, J.; Abdusalomov, A.B.; Whangbo, T.K. Attention 3D U-Net with Multiple Skip Connections for Segmentation of Brain Tumor Images. *Sensors* **2022**, *22*, 6501. [CrossRef]
38. Abdusalomov, A.B.; Mukhiddinov, M.; Whangbo, T.K. Brain Tumor Detection Based on Deep Learning Approaches and Magnetic Resonance Imaging. *Cancers* **2023**, *15*, 4172. [CrossRef]
39. Wafa, R.; Khan, M.Q.; Malik, F.; Abdusalomov, A.B.; Cho, Y.I.; Odarchenko, R. The Impact of Agile Methodology on Project Success, with a Moderating Role of Person's Job Fit in the IT Industry of Pakistan. *Appl. Sci.* **2022**, *12*, 10698. [CrossRef]
40. Norkobil Saydirasulovich, S.; Abdusalomov, A.; Jamil, M.K.; Nasimov, R.; Kozhamzharova, D.; Cho, Y.-I. A YOLOv6-Based Improved Fire Detection Approach for Smart City Environments. *Sensors* **2023**, *23*, 3161. [CrossRef]
41. Mamieva, D.; Abdusalomov, A.B.; Kutlimuratov, A.; Muminov, B.; Whangbo, T.K. Multimodal Emotion Detection via Attention-Based Fusion of Extracted Facial and Speech Features. *Sensors* **2023**, *23*, 5475. [CrossRef] [PubMed]
42. Safarov, F.; Akhmedov, F.; Abdusalomov, A.B.; Nasimov, R.; Cho, Y.I. Real-Time Deep Learning-Based Drowsiness Detection: Leveraging Computer-Vision and Eye-Blink Analyses for Enhanced Road Safety. *Sensors* **2023**, *23*, 6459. [CrossRef] [PubMed]
43. Avazov, K.; Jamil, M.K.; Muminov, B.; Abdusalomov, A.B.; Cho, Y.-I. Fire Detection and Notification Method in Ship Areas Using Deep Learning and Computer Vision Approaches. *Sensors* **2023**, *23*, 7078. [CrossRef] [PubMed]
44. Available online: https://www.kaggle.com/datasets/nigoraxonnasimova/synthetic-diabetes-2-type-prediction-dataset (accessed on 28 May 2023).

bioengineering

MDPI

Article

Implementing a Novel Machine Learning System for Nutrition Education in Diabetes Mellitus Nutritional Clinic: Predicting 1-Year Blood Glucose Control

Mei-Yuan Liu [1,2,3], Chung-Feng Liu [4,*], Tzu-Chi Lin [5,*] and Yu-Shan Ma [4]

1 Department of Nutrition, Chi Mei Medical Center, Tainan 710402, Taiwan; m880419@mail.chimei.org.tw
2 Department of Nutrition and Health Sciences, Chia Nan University of Pharmacy & Science, Tainan 710402, Taiwan
3 Department of Food Nutrition, Chung Hwa University of Medical Technology, Tainan 710402, Taiwan
4 Department of Medical Research, Chi Mei Medical Center, Tainan 710402, Taiwan; yushan.ma.72@gmail.com
5 Nursing Department, Chi Mei Medical Center, Liouying, Tainan 73657, Taiwan
* Correspondence: chungfengliu@gmail.com (C.-F.L.); clh30006@mail.chimei.org.tw (T.-C.L.)

Abstract: (1) Background: Persistent hyperglycemia in diabetes mellitus (DM) increases the risk of death and causes cardiovascular disease (CVD), resulting in significant social and economic costs. This study used a machine learning (ML) technique to build prediction models with the factors of lifestyle, medication compliance, and self-control in eating habits and then implemented a predictive system based on the best model to forecast whether blood glucose can be well-controlled within 1 year in diabetic patients attending a DM nutritional clinic. (2) Methods: Data were collected from outpatients aged 20 years or older with type 2 DM who received nutrition education in Chi Mei Medical Center. Multiple ML algorithms were used to build the predictive models. (3) Results: The predictive models achieved accuracies ranging from 0.611 to 0.690. The XGBoost model with the highest area under the curve (AUC) of 0.738 was regarded as the best and used for the predictive system implementation. SHAP analysis was performed to interpret the feature importance in the best model. The predictive system, evaluated by dietitians, received positive feedback as a beneficial tool for diabetes nutrition consultations. (4) Conclusions: The ML prediction model provides a promising approach for diabetes nutrition consultations to maintain good long-term blood glucose control, reduce diabetes-related complications, and enhance the quality of medical care.

Keywords: diabetes mellitus (DM); machine learning; artificial intelligence; feature importance; predictive system; glycosylated hemoglobin (HbA1c); well-controlled HbA1c; diabetes-related disease; nutrition education

Citation: Liu, M.-Y.; Liu, C.-F.; Lin, T.-C.; Ma, Y.-S. Implementing a Novel Machine Learning System for Nutrition Education in Diabetes Mellitus Nutritional Clinic: Predicting 1-Year Blood Glucose Control. *Bioengineering* **2023**, *10*, 1139. https://doi.org/10.3390/bioengineering10101139

Academic Editors: Hongqing Yu, Alaa AlZoubi, Yifan Zhao and Hongbo Du

Received: 24 August 2023
Revised: 24 September 2023
Accepted: 26 September 2023
Published: 28 September 2023

1. Introduction

Type 2 diabetes mellitus (T2DM) is a significant public health concern, placing a substantial burden on human life and health. It not only affects an individual's quality of life but also increases the risk of mortality and complications such as cardiovascular disease, cerebrovascular disease, diabetic nephropathy, retinopathy-induced blindness, and peripheral vascular neuropathy leading to amputation. These complications impose substantial social and economic costs [1]. Managing T2DM requires ongoing interventions, including nutritional therapy, exercise routines, medication management, self-care practices, psychological support, and smoking cessation [2]. Nutrition education plays a crucial role in the long-term management of diabetes, involving discussions, assessments, lifestyle adjustments, and ongoing monitoring for complications [3]. With guidance from a medical team, lifestyle changes and self-care knowledge taught by educators can contribute to improved prognosis, health conditions, and quality of life for patients [4].

HbA1c (glycated hemoglobin) reflects an individual's blood sugar fluctuations over the past three months before measurement and serves as an essential predictor of diabetes complications. It helps assess whether patients and their treatment are achieving or maintaining glycemic control goals [5]. HbA1c control within the first year of diabetes diagnosis strongly correlates with the occurrence of major and minor vascular diseases and mortality ten years later [6]. Accurately predicting whether a patient's HbA1c level can be less than 7% (well-controlled) within one year after the primary diagnosis can greatly assist in tailoring a long-term nutritional care plan for the patient. This approach aligns with the principles of personalized medicine and precision medicine advocated in recent years [7]. However, currently, no available tool offers personalized and accurate long-term predictions for diabetes. Recent advancements in machine learning (ML) algorithms and computing speed present an opportunity to address this gap using artificial intelligence (AI)/ML technology.

In Taiwan, the National Health Insurance Administration (NHIA) implemented a regulation in 1995 that facilitated the rapid sharing of medical information across hospitals. In 2001, the government introduced the pay-for-performance program [8], which enables the systematic monitoring and treatment of diabetic patients over an extended period. Chi Mei Medical Center, as one of Taiwan's largest hospitals, has accumulated extensive data on diabetes treatment over the past 13 years, including comprehensive records of dietitian interventions and outcomes.

In this study, we leveraged this big medical data to develop an AI system that predicts whether HbA1c levels can be well-controlled below 7% within a year after the initial diabetes diagnosis because an HbA1c level with a value of 7% is regarded as a well-controlled HbA1c level in practice [5,9]. We identified feature variables based on the medical literature and expert clinical experience. AI models often have complex nonlinear or network structures, presenting challenges in terms of interpretability. That is, explainable AI (XAI) is needed during AI development [10]. To address this, we utilized SHAP (SHapley Additive exPlanations) analysis [11], a method of XAI, to visually demonstrate the importance of each feature variable in the built prediction model.

Our AI prediction system empowers clinical health educators to understand and predict changes in HbA1c levels for diabetic patients based on their current physiological statuses. It serves as a valuable reference for clinical care and nutrition education interventions, enhancing patients' disease awareness, reinforcing the importance of lifestyle changes, and motivating positive behavioral modifications. Furthermore, by considering the AI prediction results, medical teams can intervene early; supplement their advice regarding medications, disease-specific diets, and exercise requirements; promote shared decision making; and improve the quality of medical care.

In the past, most AI/ML studies have centered on evaluating model quality, with only a limited number of predictive systems developed for medical condition prognosis [12–14]. Furthermore, predictive systems in the realm of nutrition and healthcare AI remain sparse [15]. Consequently, this study significantly adds to the advancement in this domain.

2. Materials and Methods

2.1. Research Design

In this research, we sought to develop AI models that predict whether an individual outpatient with T2DM can maintain HbA1c levels below 7% within a year of their initial diabetes diagnosis. We identified feature variables based on the medical literature and expert clinical experience. This retrospective study received approval from the institutional review board of Chi Mei Medical Center (no. 10901-014). To ensure the protection of patient privacy, all patient data were de-identified. As this study is retrospective, the need for informed consent from the patients was waived. The flowchart outlining the study process is presented in Figure 1.

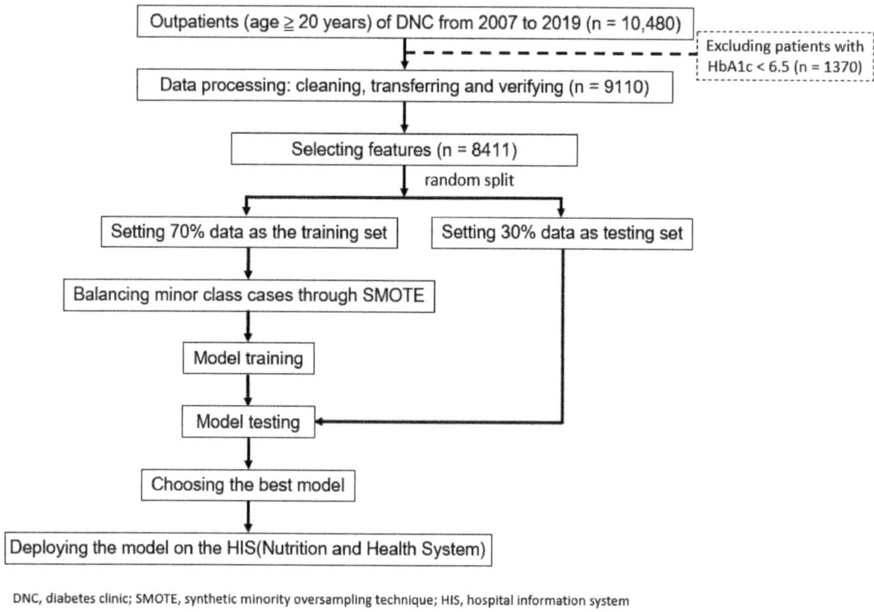

Figure 1. Research flowchart.

2.2. Setting

The data for this study were obtained from the Nutrition Education System Database of Chi Mei Medical Center in Taiwan. This study included T2DM outpatients, including those with gestational diabetes and those aged 20 years and older who participated in the pay-for-performance (P4P) program and received health education in the diabetes nutrition clinic from 2007 to the end of 2019. We ensured that there was no selective inclusion of participants, thus maintaining fairness and avoiding selection bias. Patients with a current HbA1c level of below 6.5% were excluded from this study. A total of 8411 patients were enrolled.

2.3. Definition of the Model's Outcome Variable

Maintaining HbA1c levels below 7% is clinically regarded as well-controlled blood glucose in DM patients [5,9]. Thus, we decided to set the cutoff threshold at 7% as the target to predict, with the binary outcome variable coded '1' for maintaining HbA1c levels below 7% or less after one year, and coded '0', otherwise. Patients whose current HbA1c levels were below 6.5, indicating not being diagnosed as DM, were excluded.

2.4. Feature Variables and Selection

A total of 18 feature variables, or impact factors, were proposed based on the relevant medical literature [16–20] and expert clinical experience. These variables included demographic information (age, gender, BMI, and length of illness), physical activity (exercise or no exercise), dietary intake (daily calories, average meals per day, protein, lipids, and carbohydrates), and blood biochemistry values (fasting blood glucose (glucose AC), HbA1c, total cholesterol, triglycerides (TGs), LDL cholesterol, HDL cholesterol, C-reactive protein (CRP), and estimated glomerular filtration rate (eGFR)). The feature "length of illness" denoted the duration for which a patient had been afflicted with diabetes prior to their first visit to our diabetes outpatient clinic.

2.5. Data Preprocessing and Machine Learning Modeling

The required data were extracted from the outpatient diabetic nutrition counseling system, and data with ambiguous values were checked and corrected. We observed that the pattern of missing data was consistent and appeared to be random, with each feature having a missing ratio of less than 4%. Thus, we opted to exclude the missing data without resorting to any imputation techniques. The dataset was divided into a training set (70% of the data) and a validation set (30% of the data) for model training and evaluation, respectively. Accuracy, sensitivity, specificity, and the area under the curve (AUC) were used as evaluation metrics. Prior to the model training, the training set underwent preprocessing to address data imbalances in the positive outcome using the synthetic minority over-sampling technique (SMOTE) [21]. Five supervised machine learning algorithms, including logistic regression (LR), random forest (RF), multilayer perceptron (MLP), light gradient boosting machine (Light GBM), and extreme gradient boosting (XGBoost), were used to build the models.

2.6. Prediction System Implementation and Trial Use

The best model was determined based on the AUC values, and the information technology engineers implemented the model into a prediction system for trial use by dietitians. The model was built using the Python programming language with the scikit-learn machine learning library, while the web-based user interface was created using MS Visual Studio®software (v 17.7). Both components were then integrated into an AI prediction system aimed at supporting nutrition education.

3. Results

3.1. Basic Case Information and Lifestyle Analysis

After excluding missing values, a total of 8411 patients from the diabetes nutrition clinic (DNC) at Chi Mei Medical Center were included in the machine learning model.

An analysis of basic information and daily living habits revealed 3171 patients with HbA1C levels below 7% within one year (37.7%) of their first visit, and 5240 otherwise (37.7%). There were significantly higher trends in age and average meals per day in the <7% group, indicating that older patients had a greater chance of maintaining their HbA1c levels after one year. Meanwhile, in comparison, patients with HbA1c levels greater than 7% after one year exhibited longer lengths of illness and significantly lower trends in exercise, cho. total, TG, glucose AC, and current HbA1c levels. The features of gender, BMI, protein, lipids, carbohydrates, daily calories, cho. LDL, cho. HDL, eGFR, and CRP/hs-CRP did not show significant differences between the two groups. The details are summarized in Table 1.

Table 1. Demographics and feature significance.

Variable	Overall (n = 8411)	One Year Later, HbA1c Level is Greater Than or Equal to 7 (62.3%, n = 5240)	One Year Later, HbA1c Level is Less Than 7 (37.7%, n = 3171)	*p*-Value *
Length of illness, mean (SD)	7.1 (7.3)	8.2 (7.4)	5.2 (6.7)	<0.001
Age, mean (SD)	59.5 (12.2)	59.1 (12.1)	60.0 (12.2)	0.001
Gender, n (%)				
Female	3933 (46.8)	2466 (47.1)	1467 (46.3)	0.491
Male	4478 (53.2)	2774 (52.9)	1704 (53.7)	
BMI, mean (SD)	26.0 (4.3)	26.1 (4.4)	25.9 (4.2)	0.100
Exercise, n (%)				
No	4098 (48.7)	2662 (50.8)	1436 (45.3)	<0.001
Yes	4313 (51.3)	2578 (49.2)	1735 (54.7)	

Table 1. *Cont.*

Variable	Overall (n = 8411)	One Year Later, HbA1c Level is Greater Than or Equal to 7 (62.3%, n = 5240)	One Year Later, HbA1c Level is Less Than 7 (37.7%, n = 3171)	*p*-Value *
Average meals per day, mean (SD)	2.7 (1.1)	2.7 (1.2)	2.8 (1.1)	0.004
Protein (g), mean (SD)	61.9 (23.8)	62.0 (21.7)	61.7 (26.8)	0.669
Lipids (g), mean (SD)	61.4 (26.6)	61.7 (24.8)	60.9 (29.3)	0.206
Carbohydrates (g), mean (SD)	191.0 (62.8)	190.7 (62.1)	191.4 (64.0)	0.652
Daily calories, mean (SD)	1602.1 (559.4)	1607.9 (554.1)	1592.5 (568.2)	0.224
Cho. total, mean (SD)	187.9 (47.3)	189.5 (47.6)	185.3 (46.7)	<0.001
TG, mean (SD)	169.0 (171.9)	174.9 (174.2)	159.3 (167.5)	<0.001
Cho. LDL, mean (SD)	114.0 (39.7)	114.6 (39.9)	112.9 (39.4)	0.056
Cho. HDL, mean (SD)	47.6 (13.5)	47.8 (13.7)	47.3 (13.2)	0.136
eGFR, mean (SD)	72.8 (22.8)	72.9 (23.4)	72.7 (21.8)	0.685
CRP/hs-CRP_group, n (%)				
<1	6698 (79.6)	4148 (79.2)	2550 (80.4)	0.335
$1 \leq CRP \leq 10$	772 (9.2)	487 (9.3)	285 (9.0)	
>10	941 (11.2)	605 (11.5)	336 (10.6)	
Glucose AC, mean (SD)	167.1 (70.8)	175.3 (72.4)	153.5 (66.0)	<0.001
HbA1c, mean (SD)	9.0 (2.3)	9.3 (2.2)	8.6 (2.3)	<0.001

Note: * For an alpha level of 0.05, categorical variables (gender, exercise, and CRP/hs-CRP) were evaluated using the chi-squared test approach, whereas numerical features were assessed using the two-sample t-test approach.

3.2. *Analysis of Blood Biochemistry Results*

As shown in Table 1, the blood biochemical values determined were 167.1 ± 70.8 mg/dL for fasting blood glucose, 114.0 ± 39.7 mg/dL for LDL cholesterol, 47.6 ± 13.5 mg/dL for HDL cholesterol, 187.9 ± 47.3 mg/dL for total cholesterol, 169.0 ± 171.9 mg/dL for TGs, 9.0 ± 2.3% for current glycosylated hemoglobin (HbA1c), and 72.8 ± 22.8 for the estimated glomerular filtration rate (e-GFR). Moreover, Spearman's correlation analysis identified the correlation between the outcome and each feature variable (Table 2). It revealed that length of illness and current HbA1c levels had the highest correlations with the outcome, while gender, carbohydrates, and eGFR had the lowest correlations.

Table 2. Spearman's correlations between each feature and outcome (1-year HbA1c levels < 7).

Feature	Correlation Coefficient
Length of illness	−0.244
Age	0.031
Gender	0.008
BMI	−0.016
Exercise	0.053
Average meals per day	0.022
Protein (g)	−0.010
Lipids (g)	−0.016
Carbohydrates (g)	0.005
Daily calories	−0.011
Cho. total	−0.047
TG	−0.059

Table 2. *Cont.*

Feature	Correlation Coefficient
Cho. LDL	−0.021
Cho. HDL	−0.011
eGFR	0.002
CRP/hs-CRP_group	−0.016
Glucose AC	−0.198
HbA1c	−0.215

3.3. Prediction Model Building and Feature Importance Analysis

In this study, several common and advanced machine learning algorithms were employed to predict whether patients would control HbA1c levels below 7% after one year using the 18 feature variables. The algorithms used included logistic regression (LR), random forest (RF), multilayer perceptron (MLP), light gradient boosting machine (light GBM), and extreme gradient boosting (XGBoost). A grid search with five-fold cross-validation for hyperparameter (Table 3) tuning for each algorithm was conducted to obtain the optimal model.

Table 3. Hyperparameter range for experiments.

Method and Hyperparameter	Value
XGBoost	
learning_rate	1e-3, 1e-2, 1e-1
gamma	0, 1e-2, 1e-3, 1e-4, 1e-5
n_estimators	200, 500, 750, 900, 1000
max_depth	3, 15, 25, 30, 50
num_parallel_tree	2, 5, 15
random_state	8, 16, 29, 42
objective	binary:logistic
LightGBM	
learning_rate	1e-3, 1e-2, 1e-1
n_estimators	120, 200, 500, 750, 1000
max_depth	7, 9, 15, 30, 50, 100
random_state	8, 16, 30, 42
Random forest	
n_estimators	110, 250, 500, 750, 950, 1000
max_depth	7, 9, 15, 30, 45, 50, 100
min_samples_split	2, 5, 10, 15
max_features	auto, sqrt, 0.5, 1.0, 1.5, 2.5
random_state	8, 16, 30, 42
MLP	
hidden_layer_sizes	(125), (125, 35), (100, 75, 30), (100, 55), (100, 75), (100, 45), (100), (96), (90, 60), (90)
max_iter	1000, 500, 250, 200, 100, 50, 30
learning_rate_init	1e-3, 1e-2, 1e-1
early_stopping	True, False
Logistic regression	
penalty	l1, l2
C	np.logspace (−3, 3, 7), 1, 5, 10
max_iter	7, 9, 10, 15, 50, 75, 100

Note: The hyperparameters that are not described in this table are set to the default values used in the scikit-learn library.

The accuracy of the prediction methods ranged from 0.611 to 0.690. Among these algorithms, XGBoost demonstrated the highest accuracy of 0.690, sensitivity of 0.684, specificity of 0.693, and an area under the curve (AUC) value of 0.738. The sensitivity, specificity, and AUC values for all the algorithms are presented in Table 4.

Table 4. Performance comparison of the machine learning methods (using the XGBoost model as a basis).

Algorithm	Accuracy	Sensitivity	Specificity	AUC	*p*-Value
XGBoost	0.690	0.684	0.693	0.738	-
LightGBM	0.682	0.682	0.682	0.735	0.097
Random forest	0.670	0.670	0.670	0.724	<0.001
MLP	0.633	0.632	0.633	0.667	<0.001
Logistic regression	0.611	0.611	0.611	0.634	<0.001

Note. (1) The DeLong test was utilized for significance testing. (2) The LightGBM model does not exhibit significant differences compared with the XGBoost model, whereas notable differences are observed with other models, with the XGBoost model demonstrating superior quality.

To visualize the results, the receiver operating characteristic (ROC) curves and the precision–recall curves were plotted, as shown in Figure 2 These curves provide graphical representations of the performances of the prediction models and their abilities to discriminate between positive and negative outcomes. Overall, the XGBoost algorithm was identified as the best prediction model in terms of accuracy, sensitivity, specificity, and AUC. The ROC curves and precision–recall curves provide additional insights into the performances of the models and their potential usefulness in predicting HbA1c reduction after one year. We performed the DelongTest to compare the model qualities. The results show that there was no significant difference between the LightGBM model and the XGBoost model, but the remaining models were significantly different from the XGBoost model.

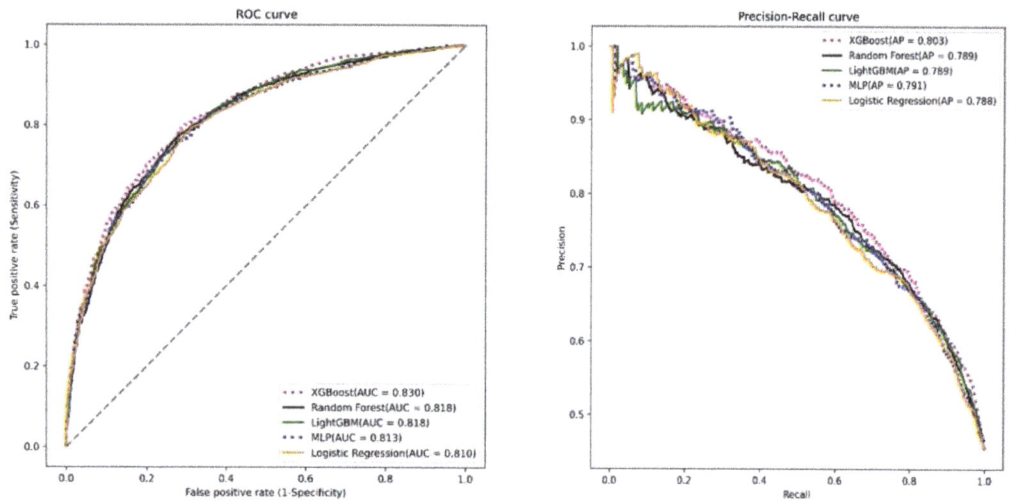

Figure 2. The receiver operating characteristic curves and the precision–recall curve.

Furthermore, we conducted a SHAP analysis for feature importance to interpret how each feature contributed to the prediction in a visual manner. A SHAP value of >0 means that it is positively related to the outcome, and vice versa. For example, in Figure 3a, the smaller the length of illness one year later, the higher the probability of controlling HBA1 below 7%, and patients with exercise habits have a higher chance of having HBA1 levels of <7 one year later. This analysis helps us understand why certain features were considered more or less important in the best XGBoost model. The feature importance plot shown in Figure 3 allows us to identify the order of importance of the model features. According to the feature importance plot of the XGBoost model (Figure 3b), we can clearly observe that the top three influential factors in the best model for predicting 1-year HbA1c levels of <7% are the length of illness, current HbA1c levels, and glucose AC.

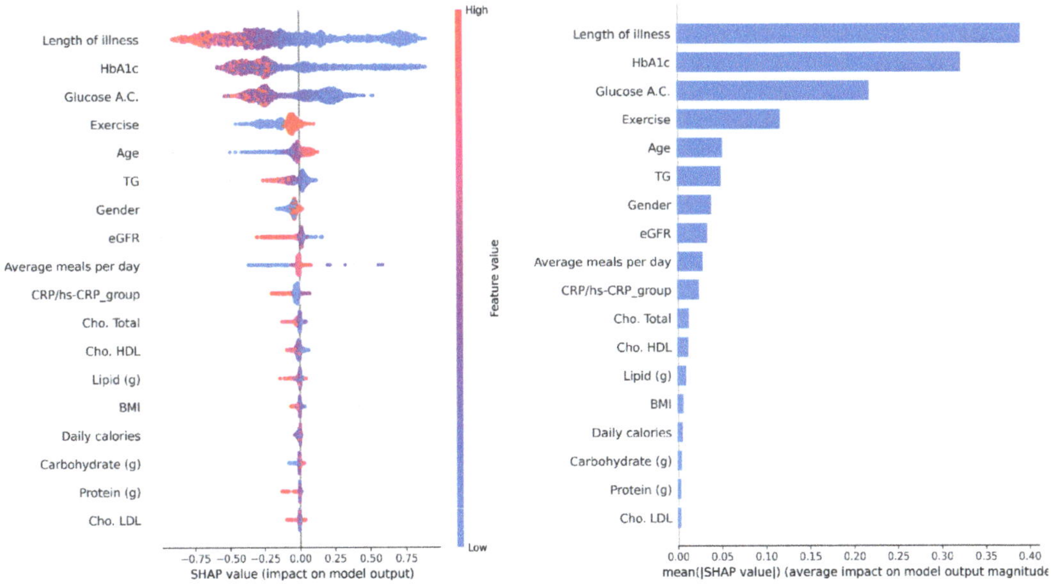

(**a**) Summary plot of SHAP values in the best model (**b**) Absolute SHAP value of each feature in the best model

Figure 3. The feature importance plots of SHAP analysis.

3.4. Prediction System Implementation and User's Acceptance

The best model was successfully implemented in a web-based forecasting system. The system screen, as depicted in Figure 4, displays the graphical interface and user-friendly design of the prediction system. This visual representation of predictions adds value by providing a clear and intuitive understanding of the patient's expected outcomes. At present, the AI prediction system has been integrated into the workflow of dietitians and provides real time and automatic prediction without manual input. Overall, the feedback received from the dietitians indicates positive acceptance and appreciation of the prediction system. The system's graphical interface and specific prediction rates were identified as valuable tools for personalized patient care and effective communication within the medical team.

Seven nutritionists were given the opportunity to use the system and provide feedback. We collected and analyzed their experiences and suggestions to assess user acceptance of the system. We asked three structured questions (on a five-point scale, one point indicating strongly disagree, and five points strongly agree): (1) Is it easy to operate? (2) Is it clinically useful? (3) Are you willing to use it? They were also encouraged to provide other comments. The survey results show that they were positive about the prediction system (the mean values of ease-of-use, usefulness, and use intention were 4.4, 3.9, and 4.1, respectively), but the score for usefulness was only 3.9, showing that the nutritionists were still not very satisfied with the system's functions. Moreover, they expressed particular appreciation for the graphical interface, which provides specific prediction rates that allow for personalized and accurate predictions of potential improvement in a patient's condition. The dietitians found that the tool could improve communication between the healthcare team and patients, facilitating discussions about subsequent nutritional treatment plans.

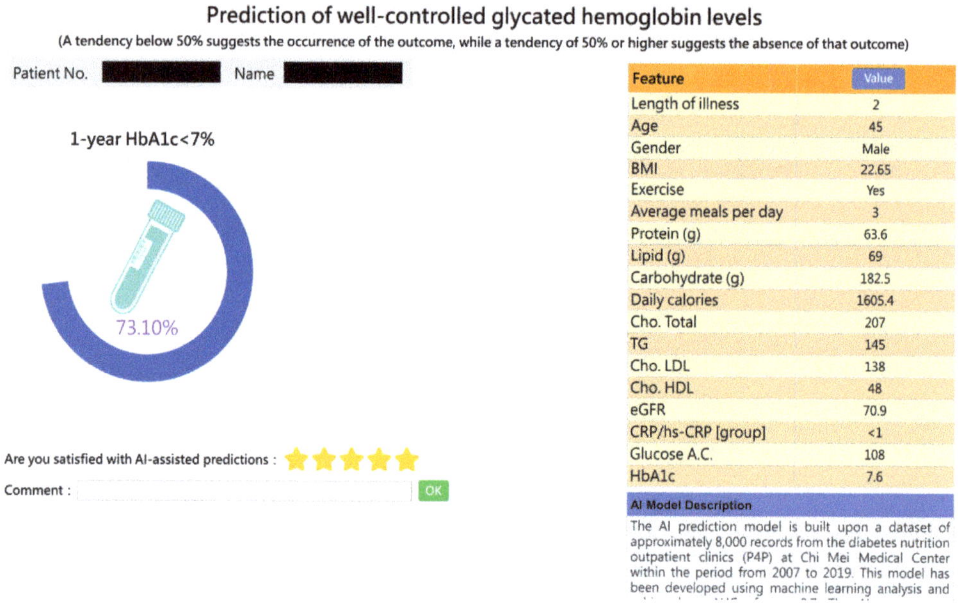

Figure 4. Prediction system picture.

3.5. Comparative Models Utilizing Alternative Feature Selection Methods

We conducted a comparison between the best all-feature model (18 features) and the significant-feature model (8 features, as indicated in Table 1) using the DeLong test. As illustrated in Table 5, although the all-feature model exhibited a slightly better performance compared with the significant-feature model, the difference did not reach statistical significance ($p = 0.085$). This suggests that the significant-feature model could be considered for clinical use, particularly when healthcare resources are limited.

Table 5. Performance comparison between all-feature model and significant-feature model.

Model	Accuracy	Sensitivity	Specificity	AUC	DeLong Test (p-Value)
All-feature model (18 features)	0.690	0.684	0.693	0.738	-
Significant-feature model (8 features)	0.678	0.679	0.678	0.734	0.058

Note: The 8 features utilized were length of illness, age, exercise, average meals per day, cho. total, TGs, glucose AC, and HbA1c.

4. Discussion

The use of AI models to develop a chronic disease nutritional status monitoring system for assessing prognostic risk is an area that lacks extensive research. However, there have been some studies exploring the use of machine learning techniques to predict the individual risk of cardiometabolic disease based on dietary or supplement intake. One such study by Panaretos et al. [22] utilized the KNN algorithm and RF decision tree to evaluate cardiometabolic risk over a 10-year period. They found that these AI/ML techniques explained a significant portion of the cardiometabolic risk, with the RF decision tree outperforming the KNN algorithm. The study also highlighted the advantages of machine learning techniques over logistic regression classification for predicting health disease risk.

The present study aligns with this research trend and contributes to it by building the best model, specifically an XGBoost-based model, which surpasses the results obtained by Panaretos et al. This study is, to the best of our knowledge, the first implementation study to utilize AI/ML technologies to predict the control of changes in HbA1c levels after one year in patients with diabetes and successfully apply it in clinical practice. By leveraging the power of AI/ML, this study expands the possibilities for personalized medicine and the use of AI in improving patient outcomes in diabetes management [23].

To explore the model's explainability, a feature importance plot was generated, revealing 12 prominent factors in the best XGBoost model. Notably, some of these factors, including current HbA1c levels, age, BMI, HDL, and eGFR, were also identified as leading factors in other models such as RF, LR, and Light GBM. This information empowers dietitians to provide targeted recommendations to patients, aiming to strengthen positive factors and mitigate negative factors, thereby increasing the likelihood of long-term reductions in HbA1c levels [24].

Based on the important features identified, we can modify them in our AI prediction system to simulate probability changes and elucidate them to patients, a process known as shared decision making (SDM). However, it is pivotal to recognize that while some elements like exercise and average meals per day can be altered through lifestyle modifications, inherent factors like age and gender remain immutable. For example, a dietitian can illustrate to a specific patient how altering the exercise feature from "No" to "Yes" can shift the probability from 45% (indicating a tendency to not achieve an HbA1c level of <7%) to 56% (indicating a tendency to achieve an HbA1c level of <7%). This visualization can motivate the incorporation of regular exercise routines, such as partaking in physical activities at least thrice a week. By concentrating on adaptable significant factors and offering tailored advice, dietitians can aid patients in effectuating substantive lifestyle modifications and enhancing long-term glycemic control.

In recent years, the digitization of medical data has revolutionized healthcare by enabling clinicians to access vast amounts of historical medical data and develop accurate predictive models for clinical decision making. This predictive tool can also be utilized by healthcare professionals to provide patients with a more precise understanding of their future outcomes, allowing them to actively participate in the decision-making process and improving communication between patients and doctors [19]. This, in turn, enhances patients' confidence in implementing the recommended changes [7].

The AI model developed in this study has been integrated into the existing DNC information system. As a result, when dietitians collect data on patients' diets, lifestyles, medication intakes, and nutritional assessments during consultations, they can utilize the predictive model seamlessly without the need for manual input. The model automatically processes the collected data to estimate the patient's HbA1c improvement one year later. This streamlined approach enables dietitians to provide timely interventions and personalized guidance on diet and lifestyle modifications, fostering effective communication between clinicians and patients in outpatient clinics [25–28].

In clinical practice, we set blood sugar control goals based on a patient's condition. Factors such as pre-meal and post-meal blood sugar levels, HbA1c values, age, and the patient's motivation to improve diabetes through lifestyle changes are all taken into account when predicting their HbA1C reduction target for the following year. With this AI prediction tool, we can assess the likelihood of achieving those goals and adjust nutritional or therapeutic plans accordingly [3]. For instance, for patients who are very likely to have their blood sugar controlled to HbA1c levels of <7% a year later (with a predicted probability of ≥50%), we intensify health education on significant features like exercise and dietary habits. We encourage them to maintain good dietary and living habits once they are back home. For patients with a tougher challenge of controlling their HbA1c levels to <7% a year later (with a predicted probability of <50%) who may struggle with consistent lifestyle and dietary habits, nutritionists not only provide active nutritional education but might also need to discuss with the attending physician about adjusting medication timings and treatment modalities. Overall, this AI prediction system serves as a smart and useful tool to achieve shared decision making between healthcare professionals and patients.

Overall, the AI system in this study stands as a pivotal tool to enhance patient awareness and motivate lifestyle alterations for optimized blood glucose control. It aids in mitigating the risks associated with both macrovascular and microvascular complications by maintaining stable glucose levels [9], ultimately serving as a facilitator in shared decision-making processes between healthcare providers and patients [29].

We recommend that practitioners integrate AI predictive models into the routine care of diabetic patients to identify high-risk individuals early and tailor interventions more effectively. Medical institutions should utilize such models to optimize resource allocation and enhance healthcare delivery, requiring proper training for practitioners in using these models. Regarding policy, it is essential to formulate and implement strategies that integrate AI technologies into healthcare protocols, advocating for the utilization of advanced technologies like IoT and wearables for real-time data acquisition and monitoring, thus improving overall disease management and mitigating the risks of complications.

5. Conclusions

In conclusion, our AI prediction system, utilizing the valuable big data accumulated at Chi Mei Medical Center, presents a novel approach for predicting a patient's 1-year HbA1c change and aiding nutritionists in making informed decisions regarding appropriate nutritional interventions. The system holds significant potential for establishing a personalized health education system, facilitating shared decision making, and enhancing the effectiveness of diabetes nutrition counseling and health education. The feature importance analysis provided a clear understanding of each feature's impact on the prediction outcome, contributing to the system's transparency and interpretability.

Furthermore, this study represents an innovative application of AI/ML technology in healthcare practice, particularly in investigating diabetic dietary habits and long-term glycemic control. It aligns with the principles of personalized precision medicine and carries substantial clinical value. We firmly believe that our prediction system can contribute to improving long-term glycemic control, reducing the incidence of diabetes-related complications, and enhancing the overall quality of medical care.

Though patients in this study were enrolled in the P4P program, the results of this study are also applicable to non-P4P patients. However, patients not enrolled in the P4P program receive fewer long-term case management follow-ups and reminders. As a consequence, their disease awareness and adherence to medical instructions may be reduced, which could subsequently impact their chances of improving their HbA1c levels.

Despite the rigorous procedure followed in this study, certain limitations should be acknowledged. Firstly, this study relied on retrospective data from a single medical center in Taiwan, potentially limiting the generalizability of the findings. Additionally, the sample was restricted to patients who participated in the P4P project, and the authenticity of nutrition counseling records, primarily relying on questions asked by medical staff and self-reported patient data, may be challenging to verify. Finally, patients' varying opinions and responses to nutrition education questions may have introduced common method bias.

Based on our results, we propose several future research directions. Firstly, the effect of medication on predictive models is an interesting but complex research topic that deserves further exploration. Secondly, gathering new patient records, referred to as a testing dataset, is valuable for estimating the expected accuracy of the proposed models to ensure generalizability. Thirdly, expanding the model's applicability and value by including patients and healthy individuals in the analysis would be beneficial. Fourthly, investigating diabetic outpatients with cardiovascular disease, cerebrovascular disease, diabetic nephropathy, and other related complications could yield valuable insights. Fifthly, considering the long-term impact of diabetes health interventions, incorporating time-series AI algorithms such as RNN and LSTM to develop long-term (multi-year) prediction models holds promise. Sixthly, prospective studies can be designed to explore patients' compliance with lifestyle changes using AI approaches [30,31]. Lastly, for real-time and continuous prediction, embracing the IoT, wearable technology, and smart technology to directly capture physiological data and daily life records (e.g., diet photos for calorie in-take determination and continuous glucose monitors (CGMs)) from patients through wearable devices and mobile apps would be a valuable avenue to pursue. However, considerations of stability, seamless connectivity, privacy, security, user-friendly interfaces, and affordability are crucial.

Author Contributions: Conceptualization: M.-Y.L. and C.-F.L.; data curation: Y.-S.M.; formal analysis: Y.-S.M. and T.-C.L.; methodology: M.-Y.L. and T.-C.L.; project administration: C.-F.L.; writing—original draft: M.-Y.L. and C.-F.L.; writing—review and editing: Y.-S.M., T.-C.L., and C.-F.L. All authors have read and agreed to the published version of the manuscript.

Funding: This research received no external funding.

Institutional Review Board Statement: This study was conducted in accordance with the Declaration of Helsinki and approved by the Institutional Review Board of Chi Mei Medical Center (protocol code: 10901-014; date of approval: 31 January 2020).

Informed Consent Statement: Patient consent was waived due to this study's retrospective nature.

Data Availability Statement: The original contributions presented in this study are included in this article. Further inquiries can be directed to the corresponding authors.

Conflicts of Interest: The authors declare no conflict of interest.

References

1. Khan, M.A.B.; Hashim, M.J.; King, J.K.; Govender, R.D.; Mustafa, H.; Al Kaabi, J. Epidemiology of Type 2 Diabetes–Global Burden of Disease and Forecasted Trends. *J. Epidemiol. Glob. Health* **2020**, *10*, 107–111. [CrossRef] [PubMed]
2. Contreras, I.; Vehi, J. Artificial Intelligence for Diabetes Management and Decision Support: Literature Review. *J. Med. Internet Res.* **2018**, *20*, e10775. [CrossRef]
3. García-Molina, L.; Lewis-Mikhael, A.M.; Riquelme-Gallego, B.; Cano-Ibáñez, N.; Oliveras-López, M.J.; Bueno-Cavanillas, A. Improving type 2 diabetes mellitus glycaemic control through lifestyle modification implementing diet intervention: A systematic review and meta-analysis. *Eur. J. Nutr.* **2020**, *59*, 1313–1328. [CrossRef] [PubMed]
4. Joachim, S.; Forkan, A.R.M.; Jayaraman, P.P.; Morshed, A.; Wickramasinghe, N. A Nudge Inspired AI-Driven Health Platform for Self-Management of Diabetes. *Sensors* **2022**, *22*, 4620. [CrossRef] [PubMed]
5. American Diabetes Association. 6. Glycemic Targets: Standards of Medical Care in Diabetes-2020. *Diabetes Care* **2020**, *43* (Suppl. S1), S66–S76. [CrossRef] [PubMed]
6. Laiteerapong, N.; Ham, S.A.; Gao, Y.; Moffet, H.H.; Liu, J.Y.; Huang, E.S.; Karter, A.J. The Legacy Effect in Type 2 Diabetes: Impact of Early Glycemic Control on Future Complications (The Diabetes & Aging Study). *Diabetes Care* **2019**, *42*, 416–426. [PubMed]
7. Hargraves, I.G.; Montori, V.M.; Brito, J.P.; Kunneman, M.; Shaw, K.; LaVecchia, C.; Wilson, M.; Walker, L.; Thorsteinsdottir, B. Purposeful SDM: A problem-based approach to caring for patients with shared decision making. *Patient Educ. Couns.* **2019**, *102*, 1786–1792. [CrossRef]
8. Hsieh, H.M.; Chiu, H.C.; Lin, Y.T.; Shin, S.J. A diabetes pay-for-performance program and the competing causes of death among cancer survivors with type 2 diabetes in Taiwan. *Int. J. Qual. Health Care* **2017**, *29*, 512–520. [CrossRef]
9. American Diabetes Association. Glycemic Targets: Standards of Medical Care in Diabetesd 2021. *Diabetes Care* **2021**, *44* (Suppl. S1), S73–S84. [CrossRef]
10. Samek, W.; Wiegand, T.; Müller, K.R. Explainable Artificial Intelligence: Understanding Visualizing and Interpreting Deep Learning Models. *arXiv* **2017**, arXiv:1708.08296.
11. Lundberg, S.M.; Lee, S.I. A Unified Approach to Interpreting Model Predictions. In Proceedings of the 31st International Conference on Neural Information Processing Systems, Red Hook, NY, USA, 4–9 December 2017.
12. Tsai, W.C.; Liu, C.F.; Ma, Y.S.; Chen, C.J.; Lin, H.J.; Hsu, C.C.; Chow, J.C.; Chien, Y.W.; Huang, C.C. Real-time artificial intelligence system for bacteremia prediction in adult febrile emergency department patients. *Int. J. Med. Inform.* **2023**, *178*, 105176. [CrossRef] [PubMed]
13. Li, Y.Y.; Wang, J.J.; Huang, S.H.; Kuo, C.L.; Chen, J.Y.; Liu, C.F.; Chu, C.C. Implementation of a machine learning application in preoperative risk assessment for hip repair surgery. *BMC Anesthesiol.* **2022**, *22*, 116. [CrossRef] [PubMed]
14. Liao, K.M.; Ko, S.C.; Liu, C.F.; Cheng, K.C.; Chen, C.M.; Sung, M.I.; Hsing, S.C.; Chen, C.J. Development of an Interactive AI System for the Optimal Timing Prediction of Successful Weaning from Mechanical Ventilation for Patients in Respiratory Care Centers. *Diagnostics* **2022**, *12*, 975. [CrossRef] [PubMed]

15. Katsimpris, A.; Brahim, A.; Rathmann, W.; Peters, A.; Strauch, K.; Flaquer, A. Prediction of type 2 diabetes mellitus based on nutrition data. *J. Nutr. Sci.* **2021**, *10*, e46. [CrossRef] [PubMed]
16. Gong, Q.H.; Kang, J.F.; Ying, Y.Y.; Li, H.; Zhang, S.H.; Wu, Y.H.; Xu, G.Z. Lifestyle interventions for adults with impaired glucose tolerance: A systematic review and meta-analysis of the effects on glycemic control. *Intern. Med.* **2015**, *54*, 303–310. [CrossRef]
17. Zhang, X.; Devlin, H.M.; Smith, B.; Imperatore, G.; Thomas, W.; Lobelo, F.; Ali, M.K.; Norris, K.; Gruss, S.; Bardenheier, B.; et al. Effect of lifestyle interventions on cardiovascular risk factors among adults without impaired glucose tolerance or diabetes: A systematic review and meta-analysis. *PLoS ONE* **2017**, *12*, e0176436. [CrossRef]
18. Jiang, Q.; Li, J.T.; Sun, P.; Wang, L.L.; Sun, L.Z.; Pang, S.G. Effects of lifestyle interventions on glucose regulation and diabetes risk in adults with impaired glucose tolerance or prediabetes: A meta-analysis. *Arch. Endocrinol. Metab.* **2022**, *66*, 157–167. [CrossRef]
19. Selya, A.; Anshutz, D.; Griese, E.; Weber, T.L.; Hsu, B.; Ward, C. Predicting unplanned medical visits among patients with diabetes: Translation from machine learning to clinical. *BMC Med. Inform. Decis. Mak.* **2021**, *21*, 111. [CrossRef]
20. Dong, Z.; Wang, Q.; Ke, Y.; Zhang, W.; Hong, Q.; Liu, C.; Liu, X.; Yang, J.; Xi, Y.; Shi, J.; et al. Prediction of 3-year risk of diabetic kidney disease using machine learning based on electronic medical records. *J. Transl. Med.* **2022**, *20*, 143. [CrossRef]
21. Chawla, N.V.; Bowyer, K.W.; Hall, L.O.; Kegelmeyer, W.P. SMOTE: Synthetic minority over-sampling technique. *J. Artif. Intell. Res.* **2002**, *16*, 321–357. [CrossRef]
22. Panaretos, D.; Koloverou, E.; Dimopoulos, A.C.; Kouli, G.M.; Vamvakari, M.; Tzavelas, G.; Pitsavos, C.; Panagiotakos, D.B. A comparison of statistical and machine-learning techniques in evaluating the association between dietary patterns and 10-year cardiometabolic risk (2002-2012): The ATTICA study. *Br. J. Nutr.* **2018**, *120*, 326–334. [CrossRef] [PubMed]
23. Afsaneh, E.; Sharifdini, A.; Ghazzaghi, H.; Ghobadi, M.Z. Recent applications of machine learning and deep learning models in the prediction, diagnosis, and management of diabetes: A comprehensive review. *Diabetol. Metab. Syndr.* **2022**, *14*, 196. [CrossRef] [PubMed]
24. Sadeghi, S.; Khalili, D.; Ramezankhani, A.; Mansournia, M.A.; Parsaeian, M. Diabetes mellitus risk prediction in the presence of class imbalance using fexible machine learning methods. *BMC Med. Inform. Decis. Mak.* **2022**, *22*, 36. [CrossRef] [PubMed]
25. Chen, Y.M.; Kao, Y.; Hsu, C.C.; Chen, C.J.; Ma, Y.S.; Shen, Y.T.; Liu, T.L.; Hsu, S.L.; Lin, H.J.; Wang, J.J.; et al. Real-time interactive artificial intelligence of things-based prediction for adverse outcomes in adult patients with pneumonia in the emergency department. *Acad. Emerg. Med.* **2021**, *28*, 1277–1285. [CrossRef]
26. Zhang, P.I.; Hsu, C.C.; Kao, Y.; Chen, C.J.; Kuo, Y.W.; Hsu, S.L.; Liu, T.L.; Lin, H.J.; Wang, J.J.; Liu, C.F.; et al. Real-time AI prediction for major adverse cardiac events in emergency department patients with chest pain. *Scand. J. Trauma Resusc. Emerg. Med.* **2020**, *28*, 93. [CrossRef]
27. Chang, Y.J.; Hung, K.C.; Wang, L.K.; Yu, C.H.; Chen, C.K.; Tay, H.T.; Wang, J.J. A Real-Time Artificial Intelligence-Assisted System to Predict Weaning from Ventilator Immediately after Lung Resection Surgery. *Int. J. Environ. Res. Public Health* **2021**, *18*, 2713. [CrossRef]
28. Lian, X.; Qi, J.; Li, X.; Wang, M.; Li, G.; Yang, T.; Zhong, J. Study on risk factors of diabetic peripheral neuropathy and establishment of a prediction model by machine learning. *BMC Med. Inform. Decis. Mak.* **2023**, *23*, 146. [CrossRef]
29. Ye, J.; Yao, L.; Shen, J.; Janarthanam, R.; Luo, Y. Predicting mortality in critically ill patients with diabetes using machine learning and clinical notes. *BMC Med. Inform. Decis. Mak.* **2020**, *20* (Suppl. S11), 295. [CrossRef]
30. Cha, E.; Clark, P.C.; Reilly, C.M.; Higgins, M.; Lobb, M.; Smith, A.L.; Dunbar, S.B. Educational needs for improving self-care in heart failure patients with diabetes. *Diabetes Educ.* **2002**, *38*, 673–684. [CrossRef]
31. Sharma, A.; Mentz, R.J.; Granger, B.B.; Heitner, J.F.; Cooper, L.B.; Banerjee, D.; Green, C.L.; Majumdar, M.D.; Eapen, Z.; Hudson, L.; et al. Utilizing mobile technologies to improve physical activity and medication adherence in patients with heart failure and diabetes mellitus: Rationale and design of the TARGET-HF-DM Trial. *Am. Heart J.* **2019**, *211*, 22–33. [CrossRef]

bioengineering

MDPI

Article

Implicit HbA1c Achieving 87% Accuracy within 90 Days in Non-Invasive Fasting Blood Glucose Measurements Using Photoplethysmography

Justin Chu [1,2], Yao-Ting Chang [3], Shien-Kuei Liaw [1] and Fu-Liang Yang [2,*]

1 Department of Electronic and Computer Engineering, National Taiwan University of Science and Technology, No. 43, Sec. 4, Keelung Rd., Taipei City 10607, Taiwan; nk95061313@gmail.com (J.C.); skliaw@mail.ntust.edu.tw (S.-K.L.)

2 Research Center for Applied Sciences, Academia Sinica, 128 Academia Rd., Sec. 2, Nankang, Taipei City 115-29, Taiwan

3 Division of Cardiology, Department of Internal Medicine, Taipei Tzu Chi Hospital, Buddhist Tzu Chi Medical Foundation, No. 289, Jianguo Rd., Xindian Dist., New Taipei City 231-42, Taiwan; necrosparkeps@tzuchi.com.tw

* Correspondence: flyang@gate.sinica.edu.tw

Abstract: To reduce the error induced by overfitting or underfitting in predicting non-invasive fasting blood glucose (NIBG) levels using photoplethysmography (PPG) data alone, we previously demonstrated that incorporating HbA1c led to a notable 10% improvement in NIBG prediction accuracy (the ratio in zone A of Clarke's error grid). However, this enhancement came at the cost of requiring an additional HbA1c measurement, thus being unfriendly to users. In this study, the enhanced HbA1c NIBG deep learning model (blood glucose level predicted from PPG and HbA1c) was trained with 1494 measurements, and we replaced the HbA1c measurement (explicit HbA1c) with "implicit HbA1c" which is reversely derived from pretested PPG and finger-pricked blood glucose levels. The implicit HbA1c is then evaluated across intervals up to 90 days since the pretest, achieving an impressive 87% accuracy, while the remaining 13% falls near the CEG zone A boundary. The implicit HbA1c approach exhibits a remarkable 16% improvement over the explicit HbA1c method by covering personal correction items automatically. This improvement not only refines the accuracy of the model but also enhances the practicality of the previously proposed model that relied on an HbA1c input. The nonparametric Wilcoxon paired test conducted on the percentage error of implicit and explicit HbA1c prediction results reveals a substantial difference, with a p-value of 2.75×10^{-7}.

Keywords: photoplethysmography; HbA1c; blood glucose

Citation: Chu, J.; Chang, Y.-T.; Liaw, S.-K.; Yang, F.-L. Implicit HbA1c Achieving 87% Accuracy within 90 Days in Non-Invasive Fasting Blood Glucose Measurements Using Photoplethysmography. *Bioengineering* **2023**, *10*, 1207. https://doi.org/10.3390/bioengineering10101207

Academic Editors: Hongqing Yu, Alaa AlZoubi, Yifan Zhao and Hongbo Du

Received: 6 September 2023
Revised: 4 October 2023
Accepted: 12 October 2023
Published: 16 October 2023

1. Introduction

1.1. Non-Invasive Blood Glucose (NIBG) Measurements

Diabetes poses a great threat to global health care. The number of people with diabetes rose from 108 million in 1980 to 422 million in 2014 [1]. More than 1.4 million cases of newly diagnosed diabetes mellitus were documented in US adults aged 18 or older in 2019 alone, and diabetes was also responsible for more than 6.7 million deaths worldwide in 2021 [2,3]. Diabetes has a prevalence of about 1 in 10 adults and causes disability and severe end-organ damage, including renal failure, retinopathy, and nerve damage, if not well controlled. The introduction of insulin and anti-diabetic medications reduces the microvascular complications of diabetes in clinical trials but also raises the concern of hypoglycemia [4]. Thus, to achieve tight glycemic control, the use of easy, convenient, point-of-care devices to supervise glucose levels is pivotal in diabetes care.

Non-invasive blood glucose (NIBG) measurement refers to the process of determining blood sugar levels without the need for traditional methods that require pricking the skin to obtain a blood sample. Traditional finger-prick measurements can cause pain and carry a risk of infection, which might discourage individuals who need to monitor their blood glucose regularly. NIBG measurement offers a more comfortable and less intrusive way to monitor glucose levels.

Numerous methods based on diverse technologies have been explored for NIBG measurement. These include enzymatic methods that test saliva, tears, and body sweat [5–7], electromagnetic wave sensing methods that cover a wide area of the electromagnetic spectrum [8–10], and transdermal methods that measure the user's bioimpedance [11].

Among these, photoplethysmography (PPG) stands out as a highly promising technology due to the fact that it is very easy to use and has very versatile applications [12]. PPG is a technology that monitors the light-absorption changes on the measured site. PPG sensors consist of a light-emitting diode (LED) that provides a stable light source, and light sensors that monitor the light intensity. While the blood volume at the measured site changes with pulsation, the light intensity also changes due to absorption and scattering. PPG allows unobtrusive continuous measurement while requiring only a single point of contact. Recent studies have even presented a breakthrough with contactless camera PPG for long-term, contactless, and continuous monitoring [13]. Various PPG-based applications are already being incorporated into commercially available products for the measurement of SpO2, stress levels, and blood pressure, and the detection of arrhythmias like atrial fibrillation.

Although the potential of relying on PPG alone to accurately estimate fasting blood glucose has frequently been explored via promising experimental results, a definitive answer remains elusive. We are of the opinion that the primary challenge stems from the missing variable or correction factor that addresses personal deviation. Every individual is inherently distinct, not only genetically but also in terms of their lifestyle and diet. Consequently, this gives rise to significant and undocumented variations among individuals when attempting to construct models for precise blood glucose level estimation.

Numerous NIBG studies based on PPG technology have been published over the past decades, employing a variety of different methods [14]. However, many of these studies suffer from small sample sizes and potentially compromise their generalizability. The most commonly employed PPG-extracted features are the morphological and heart rate variance features, which are correlated with an individual's vascular function and autonomic neuropathy [15,16]. Many studies also explore features in different domains, utilizing techniques such as fast Fourier transform (FFT), Kaiser–Teager energy (KTE), and spectral entropy [17]. It is worth noting that using an excessive number of features can lead to overfitting, while using too few features may result in a lack of vital information required for accurate blood glucose level estimation. To date, the quest for an accurate yet simple standard for a medical NIBG meter remains unfulfilled.

1.2. From Measured HbA1c to Implicit HbA1c

In our previous studies, we showcased that a universal model incorporating quarterly HbA1c measurements as input features could substantially enhance model accuracy. However, we also observed that the presence of various medications had a detrimental impact on model performance. As a result, despite the incorporation of HbA1c, our ability to generate accurate estimations remains restricted for subjects not influenced by the effects of medication [18].

HbA1c, also known as glycated hemoglobin and sometimes referred to as hemoglobin A1C, is a crucial metric used to assess long-term blood glucose control and an important indicator for diagnosing diabetes [19]. It reflects the average blood glucose level over the preceding two to three months. Over time, the glucose in the bloodstream binds itself to the hemoglobin protein. The higher the blood glucose concentration, the more glucose is bound to the protein. While HbA1c is proven to be a strong feature that can significantly enhance prediction accuracy, it is not without shortcomings. Our previously proposed HbA1c model faces the challenge of a difficulty in taking HbA1c measurement. HbA1c measurements are not as easily acquired as a finger-prick blood glucose test. HbA1c measurements are generally only available in hospitals or specialized clinics. They require more specialized equipment compared to a conventional finger-prick BGL measuring device that in comparison is already commonly available for household use. The ability to acquire a simple alternative HbA1c could significantly improve the usability of prediction models that utilize HbA1c.

To address this challenge, we herein introduce an implicit HbA1c technology aimed to achieve accurate measurement of fasting blood glucose levels using photoplethysmography alone. In Figure 1 we present a schematic depiction of the concept for this work. When working with a model that only employs PPG-extracted features to estimate blood glucose levels, sparse prediction results often emerge due to overfitting or underfitting, symbolized by the large size of the circle. To refine the circle size of the predicted blood glucose levels using the PPG feature vector, we incorporate HbA1c into the model. This addition significantly reduces the overfitting or underfitting of the dispersed BGL prediction from PPG. HbA1c is the single most meaningful variable with a significant impact on model accuracy as it is correlated to the actual BGL. It is important to clarify our terminology: in this study, we refer to the measured actual HbA1c as "explicit HbA1c" since we use this measured value from the user's result directly as it is. This is in contrast to our proposed "implicit HbA1c". Implicit HbA1c is related to explicit HbA1C but is acquired through a novel use of the same HbA1c-enhanced model to measure BGL. Table 1 showcases the straightforward nature of the implicit HbA1c method by outlining the required information input from users to operate this model. Users only need to perform a single finger-prick measurement as a pretest to derive their implicit HbA1c, which encompasses systematic correction elements that account for personal deviations, thereby further enhancing the accuracy of blood glucose level predictions.

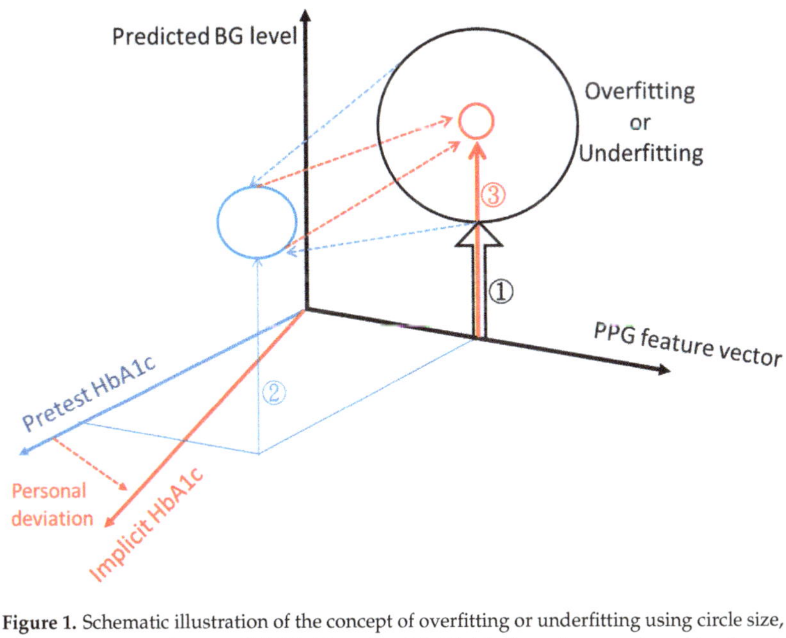

Figure 1. Schematic illustration of the concept of overfitting or underfitting using circle size, along with the transition from explicit HbA1c to implicit HbA1c for an improved machine learning prediction of blood glucose levels. To reduce the circle size of the predicted blood glucose level with the PPG feature vector, we introduce pretest HbA1c. It significantly reduces instances of overfitting or underfitting that lead to scattered blood glucose level predictions derived solely from PPG data (from ① to ②). Implicit HbA1c is related to pretest HbA1c but contains systematic correction items to account for personal deviation, thereby further enhancing the accuracy of blood glucose level prediction (from ② to ③). The input features for each stage are as follows: ① PPG features, ② PPG features and pretest HbA1c, ③ PPG features and implicit HbA1c.

Table 1. Summary of the information input and output for each section of the process for the proposed implicit HbA1c method.

	Model training	User		
		Featuring HbA1c (Pretest)	Interval up to 90 days	Testing
Required Input	• PPG signal • Reference HbA1c • Reference BGL	PPG signal Reference BGL		PPG signal Implicit HbA1c
Outcome	• BGL prediction model	Implicit HbA1c		Predicted BGL

From the user's perspective, the processes of determining implicit Hba1c are all carried out seamlessly behind the scenes. The users only need to take a single finger-prick BGL test at their initial pretest stage. The HbA1c values are all hidden from the user's view, which is why the term "implicit HbA1c" is appropriate.

2. Experiments and Method

2.1. Experimental Set-Up

From the original dataset comprising 2632 entries, a subset of 856 entries of data consisting of data from subjects not undergoing drug treatment was meticulously chosen for this study. The dataset is collected from twenty local healthcare centers across Taipei and Taoyuan County with random voluntary participants. During the data collection phase, most of the lower blood glucose subjects were unwilling to participate in the second testing, thus their data were used in model training exclusively. On the other hand, higher blood glucose subjects displayed greater enthusiasm for further testing after a few weeks to monitor changes in their blood glucose levels, as shown in Table 2. Each entry within this subset comprises two consecutive 60 s segments of PPG measurement at a 250 HZ sampling rate collected through transmissive PPG finger clips (infrared, wavelength of 940 nm) on the index finger with the TI AFE4490 Integrated Analog Front End, along with corresponding measurements of blood glucose levels via finger-pricking using the Roche Accu-Chek mobile, HbA1c using the Siemens DCA Vantage Analyzer, and blood pressure using the Omron HEM-7320. The subjects were first asked to sit on the chair in a relaxed position for at least 5 min before the measurements started. During measurement, the blood pressure and finger-prick blood glucose level measurements were taken first, immediately followed by the 60 s long PPG measurement. The collection of these samples received approval from the Institutional Review Board of the Academia Sinica, Taiwan (Application No: AS-IRB01-16081). It is noteworthy that all subjects were comprehensively informed and consented to the collection of the data and their usage.

Table 2. Characteristics of the subjects in the training and testing sets with their mean ± standard deviation.

	Dataset	Interval between Test and Pretest	BG (mg/dL)	HbA1c (%)	Age (Years)	BMI (kg/m^2)
Total 856 entries	Training (747 entries)	No pretest	99.9 ± 12.9	5.7 ± 0.53	57.9 ± 9.7	23.4 ± 3.2
	Testing (61 pairs)	45 ± 19 days	154.9 ± 50.8	7.7 ± 1.76	62.7 ± 3.95	28.4 ± 4.3

The 60 s long PPG signals are segmented into windows with a width of 400 data points (equivalent to 1.6 s) based on each pulse valley. A total of 11 features are extracted, encompassing both morphological and heart rate variance (HRV) features. The morphological features include the width of the pulse at half-height, the time taken from pulse valley to pulse peak, the sum of the pulse area of the minute, the average pulse area, and the median of the pulse area. The HRV features include the high, low, and total frequency power from fast Fourier transform (FFT), the percentage of pulse successive interval changes exceeding 20 ms, and the standard deviation of pulse successive interval changes.

In this study, our primary focus is exclusively on subjects who are not undergoing treatment with drugs. This approach serves as a follow-up to our previously proposed method, with the intent of enhancing its effectiveness. Our previous work achieved over 90% accuracy on cohorts not affected by medication with measured HbA1c employed as a feature.

For this work, subjects with multiple entries are deliberately reserved for use as the testing set, while the remaining subjects constitute the training set. The characteristics of both the training and testing sets are concisely outlined in Table 2. To align our approach with practical usage scenarios, a total of 61 pairs, each with a time interval not exceeding 90 days, are utilized for testing. Evaluating performance beyond the three-month validity of HbA1c would be both impractical and devoid of meaningful insights. Thus, the data pairs with intervals exceeding 90 days were further excluded from the testing. Within the testing set, each subject's multiple rounds of measurements are meticulously paired together in a sequential manner to establish the testing data structure. Each pair is composed of a pretest and a test measurement, collected from different measurement rounds, thereby forming varying time intervals. The valid time interval between the pretest and test spans from 11 to 90 days. Notably, none of the measurements belonging to subjects designated for the testing set are used in the training set. This meticulous separation ensures the establishment of the strictest testing condition, where the model has not yet been influenced by any prior measurements of the intended test subjects.

2.2. Method

The methodology proposed in this study revolves around the utilization of a pretest round to derive an implicit HbA1c value, subsequently enhancing the accuracy of blood glucose level (BGL) predictions during the testing round.

The workflow of this method is depicted in Figure 2. Firstly, a set of three BGL prediction models are trained employing PPG signals, PPG-extracted features, and HbA1c as inputs. These models use identical structures and will be the only prediction models used throughout the method. The three models' purpose is to validate outcomes by cross-referencing with each other. During the pretest phase, a PPG measurement with a corresponding finger-prick BGL reading is taken and inversely applied to the prediction model. The pretest PPG data are then input into the model with varying HbA1c values ranging from 4 to 12 in increments of 0.1. Consequently, a series of BGL predictions is generated, as exemplified in Figure 3. Among these predictions, the value closest to the measured finger-prick BGL reading is identified, and the corresponding HbA1c value is selected as the implicit HbA1c. This process is repeated independently for each of the three models. The disparities among the outcomes from the three models are assessed to ensure they fall within an acceptable threshold for consistency. Among the three results, the median value is chosen to serve as the designated implicit HbA1c.

During the testing phase, only the PPG measurement is collected and then joined with the pretest-determined implicit HbA1c as input for the model to generate the BGL prediction. Once more, both the PPG measurement and implicit HbA1c are independently input into the three models, and the differences among the prediction outcomes are assessed to ensure their consistency. In the event that the differences among models with identical structures and the same input data exceed a certain threshold, the results are deemed unreliable and should be disregarded. This process represents a straightforward and simple approach, leveraging preexisting models to obtain an alternative HbA1c value that enhances the accuracy of BGL predictions.

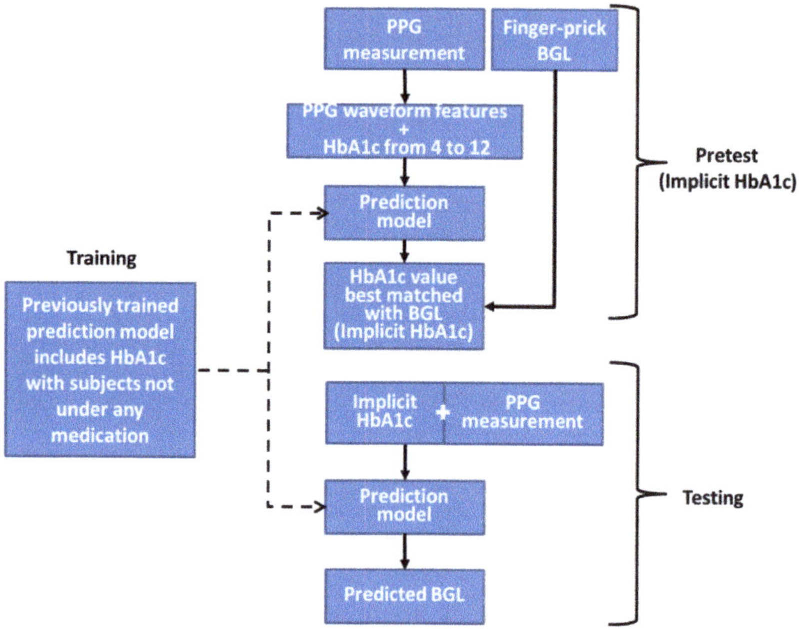

Figure 2. Illustration of the sequential workflow, starting from the pretest phase to derive the implicit HbA1c, followed by its application during testing to generate the final blood glucose level prediction. During pretest, the required inputs are PPG measurement and a finger-prick blood glucose level. The finger-prick value is only used at the final step of the pretest to determine implicit HbA1c value. During testing, the required inputs are the pretest determined implicit HbA1c and a PPG measurement.

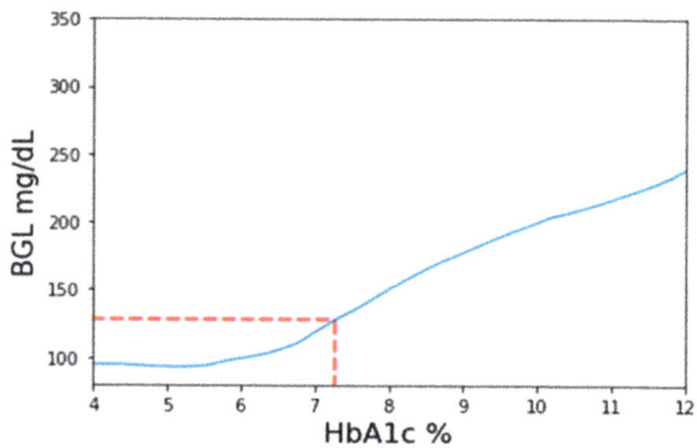

Figure 3. Illustrative example of the process for determining implicit HbA1c. A range of HbA1c values, ranging from 4 to 12, is employed to estimate blood glucose levels (BGL). Among these values, the HbA1c value corresponding to the predicted BGL closest to the pretest reference BGL is identified and designated as the implicit HbA1c value.

In this study, we utilized Python version 3.9.13 and Keras version 2.7 with tensorFlow version 2.7 as the backend for model building. The BGL prediction model utilized in this study used an identical structure to our prior HbA1c-based method, thus facilitating objective comparisons. The detailed model structure design with every layer can be found in Supplementary Figure S1. The model comprises two parallel one-dimensional convolutional neural network (CNN) blocks, each featuring different filter lengths to encompass both micro and macro perspectives of the input signal window. The CNN outputs are subsequently concatenated with a manually extracted feature vector, which includes the HbA1c measurement. This combined information is then passed through several fully connected layers to generate the BGL prediction output. A comprehensive depiction and in-depth design of the model's structure can be found in our earlier work, where the model achieved a prediction accuracy of over 94% within Clarke's error grid (CEG) zone A for subjects not influenced by any form of medication.

3. Results

In this study, Clarke's error grid (CEG) analysis is used as the main performance indicator, as the ISO 15197:2013 (International Organization for Standardization) recommendation requires personal use glucose meters to have 99% of the measurement within CEG's zones A and B [20]. Clarke's error grid analysis is a graphical method used to evaluate the accuracy of blood glucose meters developed by David Clarke in 1987 as a way to assess the clinical significance of errors in glucose measurements. CEG consists of five zones from A to E, each reflecting different clinical significance [21]. Zone A represents an accurate prediction where any differences between the prediction and reference values are considered negligible. Zone B reflects a prediction with a clinically acceptable error which could lead to unnecessary treatment but does not have a significant impact. As for Zones C to E, they represent different degrees of danger to users; if the result is used for clinical purposes, it could lead to severe harm or even death.

In Figure 4a, we presented the difference between implicit HbA1c and its corresponding measured reference HbA1c. The graph exhibits a rough alignment with the diagonal line, while implicit HbA1c values appear systematically higher than their corresponding explicit HbA1c values. Implicit HbA1c reflects the HbA1c value that the model anticipates given a specific blood glucose level. This phenomenon can be attributed to the fundamental difference between the training and testing sets we used. As a result, the training set we used (subjects without multiple entries of measurement) is predominantly composed of subjects with lower blood glucose level. In contrast, the testing set is predominantly composed of individuals with prediabetes and diabetes. Due to the methodology employed, the testing set required the test subjects to have two measurements (pretest and test), but the training set did not require a pretest for model training. This makes it impossible to mix the data between training and testing sets to achieve a more balanced distribution between the two sets. Table 2 provides a glimpse of the notable differences in average HbA1c and blood glucose levels between the training and testing datasets. From Figure 4b we can see the clear difference in the distribution of the BG–HbA1c relation between our training and testing datasets. Consequently, as we proceed with the process of calculating implicit HbA1c, the elevated fasting blood glucose levels observed within the testing subjects contribute to higher implicit HbA1c values. The systematic deviation between the explicit and implicit HbA1c values does not reflect the error; instead, it shows the amount of correction items adjusted by the methodology to bridge the population for an accurate BGL estimation.

For comparison, a set of predictions was also conducted using explicit HbA1c. This comparative analysis was carried out using the same testing dataset. The overall prediction performances by CEG's zone ratios are documented and summarized in Figure 5 shown below. In the figure, we can see the difference in prediction ability between using the newly proposed implicit HbA1c and the previously used explicit HbA1c. Overall, while using implicit HbA1c, the model not only alleviates the inconvenience associated with HbA1c measurement, but also leads to a substantial 16% improvement in prediction accuracy. This outcome is an indication that implicit HbA1c can be more effective than measured HbA1c. This intriguing phenomenon is hypothesized to stem from the implicit HbA1c calculation process which also introduced a degree of adjustment of personal deviation.

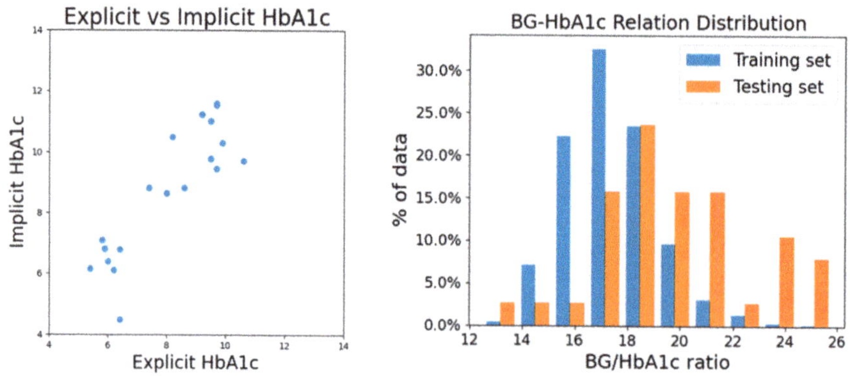

(**a**) (**b**)

Figure 4. (**a**) Difference in implicit HbA1c and corresponding measured reference HbA1c, and (**b**) distribution of the BG–HbA1c (measured) relationship for the training and testing sets.

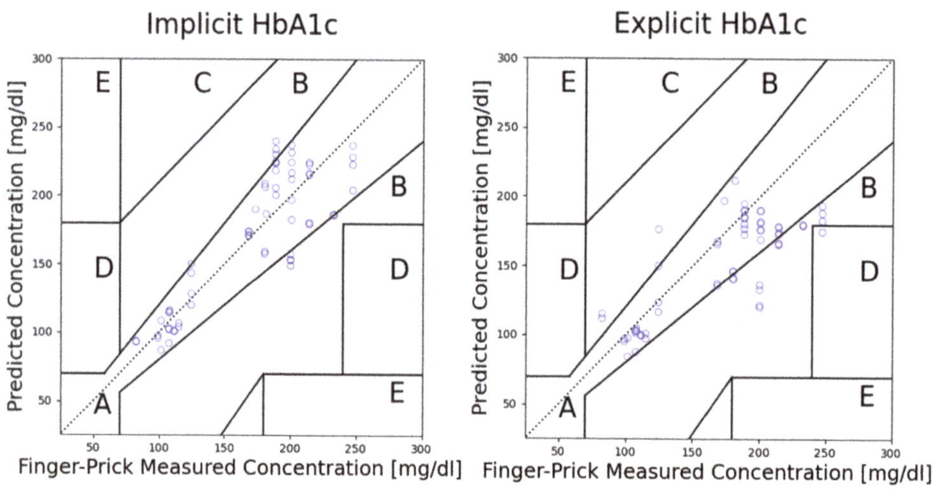

Implicit HbA1c			Explicit HbA1c		
A	B	C~E	A	B	C~E
86.9%	13.1%	0%	70.5%	27.9%	1.6%

Figure 5. CEG analysis of using implicit HbA1c and explicit HbA1c for model prediction on subjects not affected by drugs.

The distribution of the prediction percentage error of using the implicit HbA1c and explicit HbA1c methods is presented in Figure 6. In the figure, we can see the prediction percentage errors from the implicit HbA1c method exhibit a normal distribution, while those from the explicit HbA1c method displayed a left-skewed distribution with systemically lower prediction. To ascertain the significance of the difference in model performance, a nonparametric Wilcoxon signed-rank test was conducted based on the percentage errors. The statistical result revealed that there are significant differences in prediction accuracy between the two methods, with a p-value of 2.75×10^{-7}, much smaller than the significance level of 0.05.

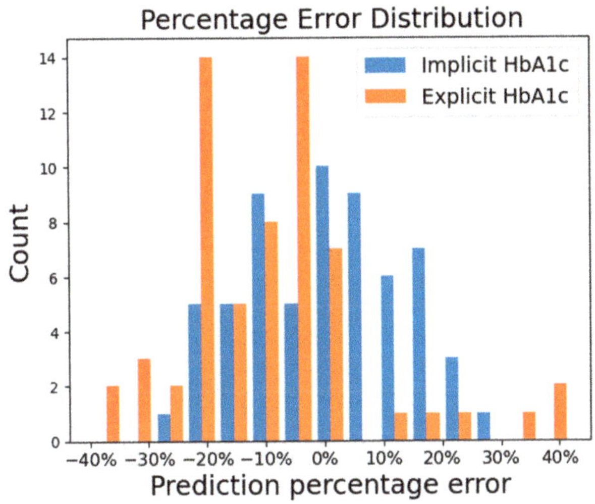

Figure 6. Percentage error distribution of the prediction while using implicit and explicit HbA1c for conducting the nonparametric Wilcoxon signed-rank test. The implicit HbA1c method demonstrates a distribution pattern that resembles a normal distribution. The explicit HbA1c method exhibits a left-skewed distribution with systemically lowered prediction.

4. Discussion

The accurate estimation of blood glucose levels (BGL) from non-medicated subjects can be achieved through a machine learning (ML) model that utilizes both photoplethysmography and HbA1c input, as we have previously demonstrated in our work published in *Sensors*. In that study, the HbA1c measurements used were taken simultaneously under the assumption that they could represent any recently measured HbA1c value with limited degradation in performance, given that HbA1c reflects a three-month average of blood glucose concentration. The less-than-ideal performance on the prediction results when using explicit HbA1c in this study was expected due to the previously mentioned disparities between the training and testing sets, as well as the increased time interval when compared to our prior work. Despite these increased challenges, the implicit HbA1c method effectively generates accurate predictions. This highlights the efficacy of implicit HbA1c in covering correction items from personal deviations.

A machine learning model for BGL estimation can generally be represented as Equation (1). Here, the function ML() symbolizes the machine learning model, while F_1 through F_n correspond to the diverse set of features that collectively contribute to achieving an accurate prediction of the blood glucose level.

$$\mathrm{BGL} = \mathrm{ML}(F_1, F_2, F_3 \ldots F_n) \tag{1}$$

While different methods may employ different features, our prior work demonstrated that BGL can be accurately estimated by a machine learning model with PPG and HbA1c input, albeit under certain conditions. This leads us to modify Equation (1) into Equation (2a):

$$\mathrm{BGL} = \mathrm{ML}(PPG, HbA1c) \tag{2a}$$

However, it is important to acknowledge the intricate interplay of variables such as medication, individual differences, lifestyle variations, and more, which may not have been fully accounted for. This realization prompts us to introduce the correction item $\sum C_i$ into the equation. For subjects not undergoing treatment with drugs, the effects of $\sum C_i$ may not be significant enough to seriously hinder the prediction performance, but it is undeniable that these effects still exist. Consequently, we further revise the equation into Equation (2b).

$$\mathrm{BGL} = \mathrm{ML}(PPG, HbA1c, \textstyle\sum C_i) \tag{2b}$$

These personal difference effects were dealt with by using a personalized deduction learning model that required multiple measurements from the user in our previous work [22]. Other works sought to account for these deviations by utilizing numerous personal profiles [23]. In this study, we leverage the concept of implicit HbA1c to achieve a similar effect.

Implicit HbA1c is determined by substituting HbA1c and BLG in Equation (2). It is like solving a multi-variate polynomial function with only one unknown variable. To solve for the unknown HbA1c value, the model is provided with a range of HbA1c inputs, generating a series of predictions. By cross-reference these predictions with the known BGL value, we can determine which corresponding HbA1c produces the most accurate estimation. This process not only yielded an HbA1c estimate, but it also accounted for the aforementioned correction items $\sum C_i$. In other words, implicit HbA1c is the HbA1c value that has been adjusted to accommodate an individual's specific correction items. Thus, this refinement further transforms Equation (2b) into Equation (3)

$$\mathrm{BGL} = \mathrm{ML}(PPG, HbA1c_{imp}) \tag{3}$$

HbA1c reflects an average BGL, and its correlation with fasting BGL is influenced by individual lifestyle, such as constant high BGL during the day and multiple meals. Consequently, the relationship between each individual's HbA1c and fasting blood glucose follows a unique curve. For instance, individuals with prediabetes may still have a pancreas capable of producing a sufficient amount of insulin to maintain normal fasting BGL, but their daily BGL may fluctuate in a big range depending on diet and lifestyle. We anticipation that the proposed method will demonstrate effectiveness across various demographics, including different races, ages, and genders, as it effectively compensates for personal deviations arising from miscellaneous correction factors.

The self-monitoring of blood glucose (SMBG) serves as an indicator of daily sugar control status in modern diabetes treatment, and its importance might be introducing behavior changes, improving glycemic control, and optimizing therapy [24]. Intensive insulin therapy is usually accompanied by daily SMBG and has proved to reduce the end-organ damage in patients with insulin-dependent diabetes mellitus [25]. There is also evidence suggesting the benefit in pre-diabetic patients or those under oral anti-diabetic drugs [26]. Some diabetes guidelines suggest SMBG use not only while fasting but also in the post-prandial stage, because the post-prandial glucose excursion, measured by the delta change in fasting and post-prandial sugar, has been demonstrated to correlate with cardiovascular risk [24]. Hence, the structured SMBG protocol by performing glucose tests before and after a meal in pairs has been evaluated in clinical trials and improves glycemic control [27]. Our implicit HbA1c method may increase the frequency of sugar monitoring compared to the guideline-suggested 2~3 times of SMBG per week in non-insulin-treated T2DM; the usage of this novel non-invasive glucose monitor technology might help diabetologists to optimize diabetic therapy in the future. However, our original dataset was collected in a fasting population; thus, the reliability of post-prandial sugar use remains uncertain. In addition, our prediction model in the insulin-treated population, whose SMBG assessments are in most demand, is less powerful than those not undergoing drug treatment. A further improvement of our algorithm and studies including a broader spectrum of diabetic populations are mandatory.

5. Conclusions

The significance of HbA1c as a valuable feature for non-invasive blood glucose prediction is widely acknowledged, although the inconvenience of acquiring HbA1c measurements remains. The HbA1c measurements are generally only available in hospitals or specialized clinics. To tackle this issue, this study introduced an innovative approach known as implicit HbA1c value which derives an alternative HbA1c that only requires a single finger-prick blood glucose measurement and can be easily conducted at home by the users. Implicit HbA1c was introduced as a solution to enable accurate glucose predictions without the need for direct HbA1c measurements with specialized equipment, and it also demonstrated the ability to further improve the prediction performance. The implicit HbA1c method achieved 87% of the prediction results within CEG's zone A, and the remaining 13% close to the zone A boundary. The implicit HbA1c approach not only exhibited a remarkable 16% improvement over the measured HbA1c method by covering personal correction items automatically, but also demonstrated an extended prediction validity period with testing data from 11 up to 90 days. The nonparametric Wilcoxon paired test conducted on the percentage error suggests a statistically significant difference between their performances with a p-value of 2.75×10^{-7}.

Supplementary Materials: The following supporting information can be downloaded at: https://www.mdpi.com/article/10.3390/bioengineering10101207/s1, Figure S1: Detailed neural network structure of our deep learning model. The model uses two input layer for signal vector and feature vector. The signal vector takes in the segmented PPG waveform data while feature vector takes in the HbA1c value and other PPG extracted features.

Author Contributions: Conceptualization, F.-L.Y.; methodology, J.C., Y.-T.C. and F.-L.Y.; writing—original draft preparation, J.C.; writing—review and editing, Y.-T.C., F.-L.Y. and S.-K.L.; supervision and project administration, F.-L.Y. and S.-K.L.; funding acquisition, F.-L.Y. All authors have read and agreed to the published version of the manuscript.

Funding: This research was supported by the Research Center for Applied Sciences of Academia Sinica, Taiwan, under grant number 3010.

Institutional Review Board Statement: The collection of samples in this study has been approved by the Institutional Review Board of Academia Sinica, Taiwan (Application No: AS-IRB01-16081). The samples were collected from a total of 2632 volunteer subjects.

Informed Consent Statement: All subjects involved in the study were fully informed and have consented to the collection of data and their use.

Data Availability Statement: Data sharing is not applicable to this article.

Acknowledgments: We thank the Ministry of Science and Technology (MOST), Taiwan, for partially supporting this study under Grant No. MOST 111-2221-E-001-005 -. We also thank the National Science and Technology Council (NSTC), Taiwan, for partially supporting this study under Grant No. NSTC 112-2221-E-001-001 -.

Conflicts of Interest: The authors declare that they have no known competing financial interest or personal relationships that could have appeared to influence the work reported in this paper.

References

1. World Health Organization. Diabetes. Available online: https://www.who.int/news-room/fact-sheets/detail/diabetes (accessed on 28 August 2023).
2. Centers for Disease Control and Prevention. National and State Diabetes Trends. Available online: https://www.cdc.gov/diabetes/library/reports/reportcard/national-state-diabetes-trends.html (accessed on 28 August 2023).
3. International Diabetes Federation. Diabetes around the World in 2021. Available online: https://diabetesatlas.org/ (accessed on 28 August 2023).
4. UK Prospective Diabetes Study (UKPDS) Group. Intensive blood-glucose control with sulphonylureas or insulin compared with conventional treatment and risk of complications in patients with type 2 diabetes (UKPDS 33). *Lancet* **1998**, *352*, 837–853. [CrossRef]
5. Aihara, M.; Kubota, N.; Minami, T.; Shirakawa, R.; Sakurai, Y.; Hayashi, T.; Iwamoto, M.; Takamoto, I.; Kubota, T.; Suzuki, R.; et al. Association between tear and blood glucose concentrations: Random intercept model adjusted with confounders in tear samples negative for occult blood. *J. Diabetes Investig.* **2021**, *12*, 266–276. [CrossRef] [PubMed]
6. Zafar, H.; Channa, A.; Jeoti, V.; Stojanovic, G.M. Comprehensive Review on Wearable Sweat-Glucose Sensors for Continuous Glucose Monitoring. *Sensors* **2022**, *22*, 638. [CrossRef] [PubMed]

7. Zhang, W.; Du, Y.; Wang, M.L. Noninvasive glucose monitoring using saliva nano-biosensor. *Sens. Bio-Sens. Res.* **2015**, *4*, 23–29. [CrossRef]
8. Monte-Moreno, E. Non-invasive estimate of blood glucose and blood pressure from a photoplethysmograph by means of machine learning techniques. *Artif. Intell. Med.* **2011**, *53*, 127–138. [CrossRef] [PubMed]
9. Rachim, V.P.; Chung, W.-Y. Wearable-band type visible-near infrared optical biosensor for non-invasive blood glucose monitoring. *Sens. Actuators B Chem.* **2019**, *286*, 173–180. [CrossRef]
10. Yadav, J.; Rani, A.; Singh, V.; Murari, B.M. Prospects and limitations of non-invasive blood glucose monitoring using near-infrared spectroscopy. *Biomed. Signal Process. Control.* **2015**, *18*, 214–227. [CrossRef]
11. Pedro, B.G.; Marcondes, D.W.C.; Bertemes-Filho, P. Analytical Model for Blood Glucose Detection Using Electrical Impedance Spectroscopy. *Sensors* **2020**, *20*, 6928. [CrossRef] [PubMed]
12. Alian, A.A.; Shelley, K.H. Photoplethysmography. *Best. Pract. Res. Clin. Anaesthesiol.* **2014**, *28*, 395–406. [CrossRef] [PubMed]
13. den Brinker, A.C.; Wang, W. Chapter 3—Model-based camera-PPG: Pulse-rate monitoring in fitness. In *Contactless Vital Signs Monitoring*; Wang, W., Wang, X., Eds.; Academic Press: Cambridge, MA, USA, 2022; pp. 51–78.
14. Hina, A.; Saadeh, W. Noninvasive Blood Glucose Monitoring Systems Using Near-Infrared Technology-A Review. *Sensors* **2022**, *22*, 4855. [CrossRef] [PubMed]
15. Benichou, T.; Pereira, B.; Mermillod, M.; Tauveron, I.; Pfabigan, D.; Maqdasy, S.; Dutheil, F. Heart rate variability in type 2 diabetes mellitus: A systematic review and meta-analysis. *PLoS ONE* **2018**, *13*, e0195166. [CrossRef] [PubMed]
16. Paneni, F.; Beckman, J.A.; Creager, M.A.; Cosentino, F. Diabetes and vascular disease: Pathophysiology, clinical consequences, and medical therapy: Part I. *Eur. Heart J.* **2013**, *34*, 2436–2443. [CrossRef] [PubMed]
17. Hina, A.; Saadeh, W. A Noninvasive Glucose Monitoring SoC Based on Single Wavelength Photoplethysmography. *IEEE Trans. Biomed. Circuits Syst.* **2020**, *14*, 504–515. [CrossRef] [PubMed]
18. Chu, J.; Yang, W.T.; Lu, W.R.; Chang, Y.T.; Hsieh, T.H.; Yang, F.L. 90% Accuracy for Photoplethysmography-Based Non-Invasive Blood Glucose Prediction by Deep Learning with Cohort Arrangement and Quarterly Measured HbA1c. *Sensors* **2021**, *21*, 7815. [CrossRef] [PubMed]
19. World Health Organization. *Use of Glycated Haemoglobin (HbA1c) in the Diagnosis of Diabetes Mellitus: Abbreviated Report of a WHO Consultation*; World Health Organization CTI—WHO Guidelines Approved by the Guidelines Review Committee: Geneva, Switzerland, 2011.
20. International Organization for Standardization. *In Vitro Diagnostic Test Systems—Requirements for Blood-Glucose Monitoring Systems for Self-Testing in Managing Diabetes Mellitus*; ISO: Geneva, Switzerland, 2013.
21. Clarke, W.L.; Cox, D.; Gonder-Frederick, L.A.; Carter, W.; Pohl, S.L. Evaluating clinical accuracy of systems for self-monitoring of blood glucose. *Diabetes Care* **1987**, *10*, 622–628. [CrossRef] [PubMed]
22. Lu, W.R.; Yang, W.T.; Chu, J.; Hsieh, T.H.; Yang, F.L. Deduction learning for precise noninvasive measurements of blood glucose with a dozen rounds of data for model training. *Sci. Rep.* **2022**, *12*, 6506. [CrossRef] [PubMed]
23. Hettiarachchi, C.; Chitraranjan, C. A Machine Learning Approach to Predict Diabetes Using Short Recorded Photoplethysmography and Physiological Characteristics. In Proceedings of the Artificial Intelligence in Medicine, Poznan, Poland, 26–29 June 2019; pp. 322–327.
24. International Diabetes Federation Clinical Guidelines Task Force. Guideline on Self-Monitoring of Blood Glucose in Non-Insulin Treated Type 2 Diabetes. *Int. Diabetes Fed.* **2009**, *17*, 4.
25. Nathan, D.M.; Genuth, S.; Lachin, J.; Cleary, P.; Crofford, O.; Davis, M.; Rand, L.; Siebert, C. The effect of intensive treatment of diabetes on the development and progression of long-term complications in insulin-dependent diabetes mellitus. *N. Engl. J. Med.* **1993**, *329*, 977–986. [CrossRef] [PubMed]
26. Sia, H.K.; Kor, C.T.; Tu, S.T.; Liao, P.Y.; Wang, J.Y. Self-monitoring of blood glucose in association with glycemic control in newly diagnosed non-insulin-treated diabetes patients: A retrospective cohort study. *Sci. Rep.* **2021**, *11*, 1176. [CrossRef] [PubMed]
27. Di Molfetta, S.; Bosi, E.; Ceriello, A.; Cucinotta, D.; Tiengo, A.; Scavini, M.; Piccolo, C.; Bonizzoni, E.; Acmet, E.; Giorgino, F.; et al. Structured self-monitoring of blood glucose is associated with more appropriate therapeutic interventions than unstructured self-monitoring: A novel analysis of data from the PRISMA trial. *Diabetes Res. Clin. Pract.* **2021**, *181*, 109070. [CrossRef] [PubMed]

MDPI
St. Alban-Anlage 66
4052 Basel
Switzerland
www.mdpi.com

Bioengineering Editorial Office
E-mail: bioengineering@mdpi.com
www.mdpi.com/journal/bioengineering

www.ingramcontent.com/pod-product-compliance
Lightning Source LLC
Chambersburg PA
CBHW061619210926
51850CB00009B/2372
9783725808038